eTELEMED 2024

The Sixteenth International Conference on eHealth, Telemedicine, and Social Medicine

May 26 - 30, 2024

Barcelona, Spain

eTELEMED 2024 Editors

Vitaly Herasevich, Mayo Clinic, USA

Jaime Lloret Mauri, Universitat Politecnica de Valencia, Spain

AA004115

Printed from e-media with permission by:

Curran Associates, Inc.
57 Morehouse Lane
Red Hook, NY 12571

Some format issues inherent in the e-media version may also appear in this print version.

Copyright© (2024) by International Academy, Research, and Industry Association (IARIA)
Please refer to the Copyright Information page.

Printed with permission by Curran Associates, Inc. (2024)

International Academy, Research, and Industry Association (IARIA)
412 Derby Way
Wilmington, DE 19810

Phone: (408) 893-6407
Fax: (408) 527-6351

petre@iaria.org

Additional copies of this publication are available from:

Curran Associates, Inc.
57 Morehouse Lane
Red Hook, NY 12571 USA
Phone: 845-758-0400
Fax: 845-758-2633
Email: curran@proceedings.com
Web: www.proceedings.com

Table of Contents

Self-Efficacy as a Determinant of Telemedicine Adoption in a Developing Country 1
Danupol Hoonsopon, Chaninun Ketkaew, Suchart Tripopsakul, Wilert Puriwat, and Wattana Viriyasitavat

A Framework for Fast Development of Customized Telehealth Applications 4
Baptiste Alcalde and Lukas Wechtitsch

A Temporal Perspective on Electronic Medicine Management Work 10
Line Lundvoll Warth and Kari Dyb

Load Induction then Simultaneous Relaxation: Insights from Multi-Modal Time-Series Data Measured with Low-Cost Wearable Sensors 16
Christoph Anders, Sai Siddhant Gadamsetti, Nico Steckhan, and Bert Arnrich

Using Computer Vision based Markerless Pose Estimation for Measuring Shoulder Range of Motion 25
Thomas Hellsten and Jonny Karlsson

A Case Study for Scoliosis: How MLOps Can Help Reduce AI Challenges in Health Care? 29
Gabor Gyorgy Gulyas, Janis Lapins, and Attila Csaba Kiss

Leveraging Voice for Early Detection of Chronic Kidney Disease: Enabling Continuous Monitoring in Remote Healthcare 37
Kangbeen Ko, Jiwon Ryu, and Sejoong Kim

Assessing Greek National Telemedicine Network 43
Haralampos Karanikas, Vasileios Tsoukas, Dimitrios Drakopoulos, George Koukoulas, Angeliki Katsapi, and Fotios Rizos

Work Related Quality of Life and HIS Usability: An Examination of Human Factors' Impact on Electronic Health Record usability during the Adoption of a New Electronic Health Record System in Norway 50
Ove Lintvedt, Espen S. Nordheim, Luis Marco-Ruiz, Terje Solvoll, and Rune Pedersen

Evaluation of Hong Kong Medical Students' Knowledge, Attitude and Intention towards the Use of Telemedicine 56
Ka Chun Fung

eTELEMED 2024

Forward

The Sixteenth International Conference on eHealth, Telemedicine, and Social Medicine (eTELEMED 2024), held between May 26th and May 30th, 2024, in Barcelona, Spain, continued a series of events considering advances in techniques, services, and applications dedicated to a global approach of eHealth.

Development of wireless homecare, of special types of communications with patient data, of videoconferencing and telepresence, and the progress in image processing and date protection increased the eHealth applications and services, and extended Internet-based patient coverage areas. Social and economic aspects as well as the integration of classical systems with the telemedicine systems are still challenging issues.

eTELEMED 2024 provided a forum where researchers were able to present recent research results and new research problems and directions related to them. The topics covered aspects from classical medicine and eHealth integration, systems and communication, devices, and applications.

We take here the opportunity to warmly thank all the members of the eTELEMED 2024 technical program committee, as well as all the reviewers. The creation of such a high-quality conference program would not have been possible without their involvement. We also kindly thank all the authors who dedicated much of their time and effort to contribute to eTELEMED 2024. We truly believe that, thanks to all these efforts, the final conference program consisted of top-quality contributions. We also thank the members of the eTELEMED 2024 organizing committee for their help in handling the logistics of this event.

We hope that eTELEMED 2024 was a successful international forum for the exchange of ideas and results between academia and industry and for the promotion of progress in the field of eHealth, telemedicine, and social medicine.

eTELEMED 2024 Chairs

eTELEMED 2024 Steering Committee
Yoshitoshi Murata, Iwate Prefectural University, Japan
Jaime Lloret Mauri, Universitat Politecnica de Valencia, Spain
Les Sztandera, Thomas Jefferson University, USA
Vitaly Herasevich, Mayo Clinic, USA

eTELEMED 2024 Publicity Chairs
Sandra Viciano Tudela, Universitat Politecnica de Valencia, Spain
Laura Garcia, Universidad Politécnica de Cartagena, Spain

eTELEMED 2024
Committee

eTELEMED 2024 Steering Committee

Yoshitoshi Murata, Iwate Prefectural University, Japan
Jaime Lloret Mauri, Universitat Politecnica de Valencia, Spain
Les Sztandera, Thomas Jefferson University, USA
Vitaly Herasevich, Mayo Clinic, USA

eTELEMED 2024 Publicity Chairs

Sandra Viciano Tudela, Universitat Politecnica de Valencia, Spain
Laura Garcia, Universidad Politécnica de Cartagena, Spain

eTELEMED 2024 Technical Program Committee

Shabbir Syed Abdul, Taipei Medical University, Taiwan
Don Adjeroh, West Virginia University, USA
Giovanni Albani, Istituto Auxologico Italiano - IRCCS, Italy
Basel Almourad, College of Technological Innovation, Dubai, UAE
Domingos Alves, Ribeirao Preto Medical School | University of Sao Paulo (USP), Brazil
Prima Oky Dicky Ardiansyah, Iwate Prefectural University, Japan
Antonio Augusto Goncalves, Estacio de Sá University, Brazil
Rafael Avila, Hospital Universitario Privado de Cordoba, Argentina
Mansoor Baig, King Faisal Specialist Hospital & Research Center, Riyadh, Saudi Arabia / ICIMTH, Greece
Panagiotis D. Bamidis, School of Medicine - Aristotle University of Thessaloniki, Greece
Oresti Baños, University of Granada, Spain
Ivana Bartoletti, Gemserv, UK
Azadeh Bashiri, Shiraz University of Medical Sciences, Iran
Hrvoje Belani, Ministry of Health - Directorate for e-Health, Zagreb, Croatia
Elisabetta Benevento, University of Pisa, Italy
Arriel Benis, HIT- Holon Institute of Technology, Israel
Sid-Ahmed Berrani, Ecole Nationale Polytechnique, Algiers, Algeria
Anna M. Bianchi, Politecnico di Milano, Italy
Vilmos Bilicki, University of Szeged, Hungary
Lucia Billeci, National Research Council of Italy | Institute of Clinical Physiology, Pisa, Italy
Antonis Billis, School of Medicine - Aristotle University of Thessaloniki, Greece
Tetiana Biloborodova, G.E. Pukhov Institute for Modelling in Energy Engineering, Ukraine
Sylvie Briand, World Health Organization, Switzerland
Taxiarchis Botsis, The Sidney Kimmel Comprehensive Cancer Center | Johns Hopkins University School of Medicine, USA
Eman Buhagiar, Middlesex University, Malta
Marco Buzzelli, University of Milano - Bicocca, Italy
Enrico Gianluca Caiani, Politecnico di Milano, Italy
Jessica Campbell, University of Central Florida / Optum, USA
Manuel Campos Martínez, University of Murcia, Spain

Nicola Carbonaro, University of Pisa, Italy
Ayan Chatterjee, The University of Agder, Grimstad, Norway
Darwyn Chern, Copa Health, Phoenix, USA
Bhargava Chinni, University of Rochester, USA
Mario Ciampi, National Research Council of Italy | Institute for High Performance Computing and Newtorking, Italy
James J. Cimino, Informatics Institute - University of Alabama at Birmingham, USA
Javier Civit, Cober SL, Spain / Gnomon Informatics, Greece
Daniel Condor Camara, Cayetano Heredia University, Peru
Massimo Conti, Università Politecnica delle Marche, Ancona, Italy
Sandra Costanzo, University of Calabria, Italy
Paul M. Cunningham, IST-Africa Institute, Ireland
Jacques Demongeot, Université Grenoble Alpes, France
Gayo Diallo, Univ. Bordeaux/ISPED, France
Linying (Lin) Dong, Ryerson University, Canada
Alexandre (Sasha) Douplik, Toronto Metropolitan University, Canada
Audrey DunnGalvin, University College Cork, Ireland
Duarte Duque, 2Ai | Polytechnic Institute of Cávado and Ave Barcelos, Portugal
Claudio Eccher, FBK Fondazione Bruno Kessler, Italy
Dina El Demellawy, CHEO Research Institute | University of Ottawa, Canada
Mohamed El Hafedh Abdi, Centre d'imagerie scintigraphique Blida, Algeria
Christo El Morr, York University, Canada
Radwa El Shawi, University of Tartu, Estonia
Manuel Filipe Santos, University of Minho, Portugal
Bruno Fionda, Fondazione Policlinico Universitario "A. Gemelli" IRCCS, Italy
Sebastian Fudickar, Universität Oldenburg, Germany
Niels F. Garmann-Johnsen, University of Agder, Norway
Anthony Gelibert, Carbon Bee, France
Wojciech Glinkowski, Polish Telemedicine Society / Center of Excellence "TeleOrto", Poland
Kuang Gong, Massachusetts General Hospital / Harvard Medical School, USA
Manuel González-Hidalgo, University of the Balearic Islands / Balearic Islands Health Research Institute (IdISBa), Spain
Conceição Granja, Norwegian Centre for e-health Research, University Hospital of North Norway, Norway
David Greenhalgh, University of Strathclyde, Glasgow, UK
Teresa Guarda, Universidad Estatal Peninsula Santa Elena - UPSE, Ecuador
Katarina Gvozdanovic, Agency for Medicinal Products and Medical Devices, Zagreb, Croatia
Mohammad Hassanzadeh, Tarbiat Modares University, Iran
Vitaly Herasevich, Mayo Clinic, USA
Pilar Herrero, Universidad Politécnica de Madrid, Spain
Felix Holl, DigiHealth Institute - Neu-Ulm University of Applied Sciences / IBE - University of Munich, Germany
Delowar Hossain, BRAC University | United International University | IDCL Fiance Limited, Bangladesh
Ying-Feng Hsu, Osaka University, Japan
Yan Hu, Blekinge Institute of Technology, Sweden
Maryam Jafarpour, University of Tehran / Ministry of Health and Medical education, Iran
Tian Kang, Tempus Labs Inc., USA
Ashad Kabir, Charles Sturt University, Australia

Haralampos Karanikas, University of Thessaly, Greece
Martijn Kiers, University of Applied Science FH JOANNEUM, Austria
Toralf Kirsten, University of Applied Sciences Mittweida, Germany
Monika Knudsen Gullslett, Nasjonalt Senter for e-Helseforskning, Sweden
Evdokimos Konstantinidis, Aristotle University of Thessaloniki, Greece / Nively, Nice, France
Stathis Th. Konstantinidis, School of Health Sciences | University of Nottingham, UK
Valencia Koomson, Tufts University, USA
Frank Kramer, Faculty of Medicine/University of Augsburg, Germany
Vinay Kumar, Thapar University, Patiala,India
Mouhamadou Lamine Ba, Ecole Supérieure Polytechnique | Université Cheikh Anta Diop, Dakar, Sénégal
Daniele Landi, University of Bergamo, Italy
Ettore Lanzarone, University of Bergamo, Italy
Edward R. Laskowski, Mayo Clinic, USA
Carla V. Leite, University of Aveiro, Portugal
Ove Lintvedt, Norwegian Centre for E-health Research / Nord University, Norway
Siru Liu, University of Utah, USA
Tatjana Lončar-Turukalo, University of Novi Sad, Serbia
Ana Rita Londral, Nova University of Lisbon, Portugal
Guillermo Lopez Campos, Wellcome-Wolfson Institute for Experimental Medicine | Queen's University Belfast, UK
Ljerka Luic, University North, Croatia
Gang Luo, University of Washington, USA
Rafael Maestre Ferriz, CETEM, Spain
Flora Malamateniou, University of Piraeus, Greece
Sadouanouan Malo, University Nazi Boni, Burkina Faso
Luis Marco-Ruiz, Norwegian Centre for E-health Research | University Hospital of North Norway, Tromsø, Norway /Peter L. Reichertz Institute for Medical Informatics of TU Braunschweig |Hannover Medical School, Germany
Giancarlo Mauri, University of Milano-Bicocca, Italy
Enkeleint-Aggelos Mechili, University of Vlora, Albania / University of Crete, Greece
Julio César Mello Román, Universidad Nacional de Asunción, Paraguay
Alessandro Mengarelli, Università Politecnica delle Marche, Ancona, Italy
Robert Mischak, Graz University of Applied Sciences, Austria
Sandra Mitrovic, IDSIA - USI/SUPSI (Dalle Molle Institute for Artificial Intelligence), Switzerland
Layal Mohtar, American University of Beirut Medical Center, Lebanon
António H. J. Moreira, 2Ai - Polytechnic Institute of Cávado and Ave, Barcelos, Portugal
Fernando Moreira, Universidade Portucalense, Portugal
Mário W. L. Moreira, Federal Institute of Education, Science, and Technology of Ceará, Brazil
Yoshitoshi Murata, Iwate Prefectural University, Japan
Sahiti Myneni, The University of Texas | School of Biomedical Informatics, USA
Paolo Napoletano, University of Milan-Bicocca, Italy
Yuriy L. Orlov, Russian Academy of Sciences | The Digital Health Institute I.M. Sechenov, Russia
Nuria Ortigosa, Universitat Politecnica de Valencia, Spain
Anna Pastusiak, StethoMe® / Adam Mickiewicz University, Poznan, Poland
Hugo Peixoto, Algoritmi Research Center | University of Minho, Portugal
Vitor Pinheiro de Almeida, Pontifícia Universidade Católica do Rio de Janeiro (PUC-Rio), Brazil
Ivan Miguel Pires, Instituto de Telecomunicações | Universidade da Beira Interior, Covilhã, Portugal
Prasad Ponnapalli, Manchester Metropolitan University, UK

Filipe Portela, University of Minho, Portugal
Sandhya Prabhakaran, Moffitt Cancer Center, Tampa, USA
Rüdiger Pryss, University of Würzburg, Germany
Ilir Qose, Aicare Srl, Italy
Taoufik Rachad, University of Mohammed V, Rabat, Morocco
M. Sohel Rahman, Bangladesh University of Engineering and Technology, Bangladesh
Gurprit K. Randhawa, First Nations Health Authority / University of Victoria / McMaster University, Canada
Sónia Rolland Sobral, Universidade Portucalense, Portugal
Carlos Rompante Cunha, CeDRI & UNIAG & Polytechnic Institute of Bragança, Portugal
Juha Röning, University of Oulu, Finland
Priscila T. M. Saito, Federal University of Technology - Parana (UTFPR), Brazil
Hayri Sever, Cankaya University, Turkey
Gro-Hilde Severinsen, Norwegian centre for e-health research, Norway
Zubair Shah, Hamad Bin Khalifa University (HBKU), Qatar
Rosa Sicilia, University Campus Bio-Medico of Rome, Italy
Line Silsand, Norwegian Centre for E-health Research, Norway
Åsa Smedberg, Stockholm University, Sweden
Sweta Sneha, Kennesaw State University, USA
Alessandro Stefanini, University of Pisa, Italy
Vasile Stoicu-Tvadar, University Politehnica Timisoara, Romania
Muhammad Sulaiman, UiT The Arctic University of Norway, Norway
Kenji Suzuki, Tokyo Institute of Technology, Japan
Alessandro Tognetti, University of Pisa, Italy
Alessandro Tonacci, Institute of Clinical Physiology | National Research Council of Italy (IFC-CNR), Pisa, Italy
Niruwan Turnbull, Mahasarakham University, Thailand
Gary Ushaw, Newcastle University, UK
Aristides Vagelatos, CTI&P, Athens, Greece
Irina Vasilyeva, The Russian State Medical University, Moscow, Russia
José Luis Vázquez Noguera, Universidad Nacional de Asunción, Paraguay
Laura Vera Righi, National Cancer Institute, Uruguay
Henrique Vicente, University of Évora, Portugal
Dongwen Wang, Arizona State University, USA
Utoomporn Wongsin, Foundation for Research Institute on Social Protection and Health, Thailand
Takashi Yamauchi, Texas A&M University, USA
Ping Yu, University of Wollongong, Australia
Zhongming Zhao, University of Texas Health Science Center at Houston, USA
Huiru (Jane) Zheng, Ulster University, UK
Stelios Zimeras, University of the Aegean, Greece
Evi Zouganeli, OsloMet - Oslo Metropolitan University, Norway
Emmanouil A. Zoulias, School of Health Sciences - National and Kapodistrian University of Athens, Greece

Copyright Information

For your reference, this is the text governing the copyright release for material published by IARIA.

The copyright release is a transfer of publication rights, which allows IARIA and its partners to drive the dissemination of the published material. This allows IARIA to give articles increased visibility via distribution, inclusion in libraries, and arrangements for submission to indexes.

I, the undersigned, declare that the article is original, and that I represent the authors of this article in the copyright release matters. If this work has been done as work-for-hire, I have obtained all necessary clearances to execute a copyright release. I hereby irrevocably transfer exclusive copyright for this material to IARIA. I give IARIA permission or reproduce the work in any media format such as, but not limited to, print, digital, or electronic. I give IARIA permission to distribute the materials without restriction to any institutions or individuals. I give IARIA permission to submit the work for inclusion in article repositories as IARIA sees fit.

I, the undersigned, declare that to the best of my knowledge, the article is does not contain libelous or otherwise unlawful contents or invading the right of privacy or infringing on a proprietary right.

Following the copyright release, any circulated version of the article must bear the copyright notice and any header and footer information that IARIA applies to the published article.

IARIA grants royalty-free permission to the authors to disseminate the work, under the above provisions, for any academic, commercial, or industrial use. IARIA grants royalty-free permission to any individuals or institutions to make the article available electronically, online, or in print.

IARIA acknowledges that rights to any algorithm, process, procedure, apparatus, or articles of manufacture remain with the authors and their employers.

I, the undersigned, understand that IARIA will not be liable, in contract, tort (including, without limitation, negligence), pre-contract or other representations (other than fraudulent misrepresentations) or otherwise in connection with the publication of my work.

Exception to the above is made for work-for-hire performed while employed by the government. In that case, copyright to the material remains with the said government. The rightful owners (authors and government entity) grant unlimited and unrestricted permission to IARIA, IARIA's contractors, and IARIA's partners to further distribute the work.

Self-Efficacy as a Determinant of Telemedicine Adoption in a Developing Country

Danupol Hoonsopon
Department of Marketing
Chulalongkorn University
Bangkok, Thailand
e-mail: danupol@cbs.chula.ac.th

Chaninun Ketkaew
Chonburi Cancer Hospital
Chonburi, Thailand
e-mail: tokkid@yahoo.com

Suchart Tripopsakul
Shool of Entrepreneurship and Management
Bangkok University
Pathumthani, Thailand
e-mail: suchart.t@bu.ac.th

Wilert Puriwat
Department of Marketing
Chulalongkorn University
Bangkok, Thailand
e-mail: wilert@cbs.chula.ac.th

Wattana Viriyasitavat
Department of Statistics
Chulalongkorn University
Bangkok, Thailand
e-mail: hardgolf@gmail.com

Abstract— **Existing literature has investigated the role of self-efficacy on the adoption of telemedicine. However, few studies explore the sources of self-efficacy. Enactive mastery, vicarious experience, and verbal persuasion have been proposed in this study. Moreover, the cultural differences may affect how individuals respond to the factors that influence their self-efficacy and adoption of telemedicine. This study expands the literature on self-efficacy's effect and on how to promote telemedicine in an emerging market.**

Keywords- ***self-efficacy; telemedicine; adoption; developing country.***

I. INTRODUCTION

Self-efficacy has influenced the behavior of individuals to do or not to do something [1] [2]. In the past, several studies [3] [4] have attempted to investigate the role of psychological factors on adopting of technology in healthcare industry. Many studies [5] - [7] agree that self-efficacy is a major factor that influences the use of telemedicine by an individual. In [1], the author defines self-efficacy as the expectations of individual efficacy which determine initiating behavior, how much effort will be dedicated, and how long he/she sustains the obstacle encountered. In [1] [8], the authors classify self-efficacy into three sources: enactive mastery, vicarious experience, and verbal persuasion.

Previous literature argues that individual beliefs such as self-efficacy causes individual action [9]. These studies mainly focus on enacting mastery, which is one source of self-efficacy. However, to the best of our knowledge, other sources of self-efficacy which are vicarious experience and verbal persuasion have been paid little attention to in previous literature.

Additionally, culture in developed countries and emerging countries is not the same [10]. Culture differences provide dissimilar reasons to adopt new products and services [11]. It is possible that the role of self-efficacy on the adoption of telemedicine may not the same in developed and emerging countries. In [6], the authors argue that many literature works examine factors that effect telemedicine adoption in developed or Western countries. A few literature works such as [7] investigate the impact of self-efficacy on telemedicine adoption in emerging country. To enhance the understanding of self-efficacy in emerging countries, this study investigates the effect of elements (enactive mastery, vicarious experience, and verbal persuasion) of self-efficacy on the adoption of telemedicine.

The rest of the paper is structured as follows. The theoretical foundation is explained in Section 2. We present our methodology in Section 3 and discussion in Section 4. Section 5 concludes the paper.

II. THEORETICAL FOUNDATION

A. Self-efficacy

Self-efficacy theory is proposed by [1], which it is defined as the expectations of individual efficacy which determine initiating behavior, how much effort will be dedicated, and how long he/she sustains the obstacle encountered. In [1] [8], they suggest that there are three sources of self-efficacy: enactive mastery, vicarious experience, and verbal persuasion. Enactive mastery refers to the degree of recognition of individuals to their ability to success on tasks [8]. Vicarious experience is defined as individuals perceive behavior of others (e.g., friends, family, influencers, and role models), observe what they are able to do, evaluate the outcome of their behavior, and use this information to be a guideline for doing something [2] [12]. Lastly, verbal persuasion is individuals who are convinced by people that have ability to success in a specific task [13].

After [1] formulated self-efficacy theory, this theory is applied into various areas which one application area is telemedicine adoption [7]. In [5], the author suggests that social cognitive factor such as self-efficacy is strongly related to healthcare app adoption. Individuals who have high self-efficacy tend to recognize their ability to use telemedicine, observe how to use telemedicine from others, and follow the suggestion from influencing people.

B. *Cultural Difference in Technology Adoption*

Cultural differences provide different norms, beliefs, attitudes and behavior of individuals in each society such as individualism vs collectivism, uncertainty avoidance, and long term versus short term orientation [10]. Existing literature investigates the impact of self-efficacy on telemedicine adoption in Western context. However, norms of Western and Eastern countries are not the same. Different norms reflect on attitude and behavior of individuals which may lead to the rate of technology adoption [14]. In collectivism, low uncertainty avoidance and long-term orientation, the impact of imitation among individuals in society dominate to technology adoption [14].

C. *The impact of self-efficacy on telemedicine adoption*

Although there are many literature works explaining the impact of self-efficacy on telemedicine adoption, these studies focus on enactive mastery dimension. Other dimensions of self-efficacy which are vicarious experience and verbal persuasion lacks of examination in telemedicine context. This study proposes the impact of three dimensions of self-efficacy on telemedicine adoption. A conceptual framework of this study is shown in Figure 1.

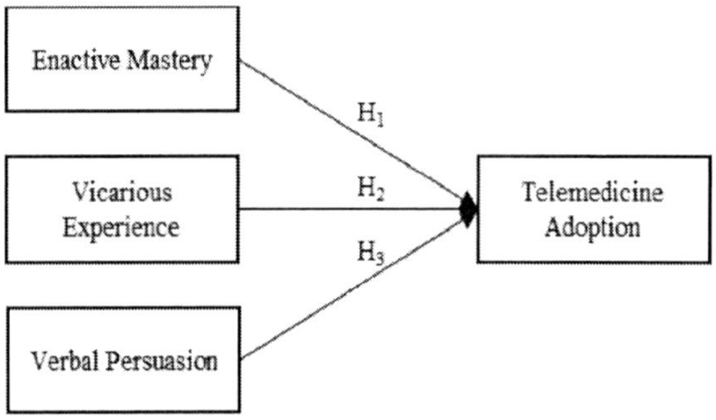

Figure 1. A conceptual framework

We propose the three sources of self-efficacy that have impacted on the telemedicine adoption which are enactive mastery, vicarious experience, and verbal persuasion as follow:

H_1: Enactive mastery has a positive impact on telemedicine adoption

H_2: Vicarious experience has a positive impact on telemedicine adoption

H_3: Verbal persuasion has a positive impact on telemedicine adoption

III. METHODOLOGY

The population of this study comprises individuals in Thailand who have experience with medical services. Many Thais have developed proficiency in using digital technology such as mobile phone and APP, particularly government support during COVID-19 pandemic crisis. We collect the data using a structured questionnaire. For data analysis and hypothesis testing, we conduct structural equation modeling with a two-step approach. Additionally, descriptive statistics are reported.

IV. DISCUSSION

Self-efficacy is a key psychological factor that can increase the adoption rate of telemedicine. This study identifies three sources of self-efficacy: enactive mastery, vicarious experience, and verbal persuasion. These sources can be used by stakeholders (e.g., policy makers, hospital industry, and academicians) to promote the adoption of telemedicine through psychological interventions. For example, we can motivate individuals to recognize their ability to use telemedicine, we can use groups as role models for individuals to follow, and we can use social influence to convey the value of telemedicine.

V. CONCLUSION

While this study sheds light on the role of self-efficacy in adopting telemedicine, it is essential to consider other critical factors in the future. These factors include technological readiness and psychological barriers related to new technology. Moreover, variations between Eastern and Western countries may impact telemedicine adoption rates. Understanding these dynamics can guide policymakers in designing effective strategies to promote telemedicine adoption, particularly in developing countries. By fostering telemedicine adoption, we aim to enhance the quality of life for people in these regions.

ACKNOWLEDGMENT

This project is partly funded by National Research Council of Thailand (NRCT), Project no. N42A660902.

REFERENCES

[1] A. Bandura, "Self-efficacy: toward a unifying theory of behavioral change," Psychological Review, vol. 84, pp. 191-215, March 1977.

[2] J. E. Maddux, "Self-efficacy theory," in Self-efficacy, adaptation, and adjustment, J. E. Maddux, Eds. Springer, Boston, MA, pp. 3-33, 1995.

[3] D. Chinn and C. McCarthy, "All Aspects of Health Literacy Scale (AAHLS): developing a tool to measure functional, communicative and critical health literacy in primary healthcare settings," Patient Education and Counseling, vol. 90, pp. 247-253, 2013.

[4] N. Sripawatakul, W. Puriwat, and D. Hoonsopon, "The impact of digital service quality toward customer engagement: A case study of telemedicine in Thailand," International Journal of Professional Business Review, vol. 8, e02231-e02231, May 2023.

[5] L. Dam, D. Roy, D. J. Atkin, and D. Rogers, "Applying an integrative technology adoption paradigm to health app adoption and use," Journal of Broadcasting & Electronic Media, vol. 62, pp. 654-672, November 2018.

[6] C. S. Kruse, P. Karem, K. Shifflett, L. Vegi, K. Ravi, and M. Brooks, "Evaluating barriers to adopting telemedicine worldwide: a systematic review," Journal of Telemedicine and Telecare, vol. 24, pp. 4-12, January 2018.

[7] X. Zhang, X. Han, Y. Dang, F. Meng, X. Guo, and J. Lin, "User acceptance of mobile health services from users' perspectives: The role of self-efficacy and response-efficacy in technology acceptance," Informatics for Health and Social Care, vol. 42, pp. 194-206, 2017.

[8] H. Margolis and P. P. McCabe, "Improving self-efficacy and motivation: What to do, what to say," Intervention in School and Clinic, vol. 41(4), pp. 218-227, March 2006

[9] V. L. Champion and C. S. Skinner, "The health belief model," in Health Behavior and Health Education: Theory, Research, and Practice, K. Glanz, B. K. Rimer, and K. Viswanath, Eds. 4th ed., Jossey-Ba, pp. 45-65, 2008.

[10] G. Hofstede, Culture's consequences: International differences in work-related values. Sage, 1984.

[11] D. Hoonsopon, "Accelerating Adoption of New Products of Thai Consumers: The Moderating Roles of Self- brand Concept and Reference Group," Journal of Asia-Pacific Business, vol. 17(2), pp. 151-172, May 2016

[12] P. Pransopon, and D. Hoonsopon, "The Impact of Reference Groups on the Purchase Intentions of Sporting Products: The Case for Spectatorship and Participation," Asian Academy of Management Journal, vol. 24(1), pp. 1-23. June 2019

[13] F. C. Lunenburg, "Self-efficacy in the Workplace: Implications for Motivation and Performance," International Journal of Management, Business, and Administration, vol. 14, pp. 1-6. 2011.

[14] S. G. Lee, S. Trimi, and C. Kim, "The Impact of Cultural Differences on Technology Adoption," Journal of World Business, vol. 48, pp. 20-29. January 2013.

A Framework for Fast Development of Customized Telehealth Applications

Baptiste Alcalde
eHealth Institut
FH JOANNEUM
Graz, Austria
baptiste.alcalde@fh-joanneum.at

Lukas Wechtitsch
Independent
Graz, Austria
wechtitsch.lukas@gmail.com

Abstract — **Telehealth has the potential to enhance health services and lower their costs, particularly in remote regions. Most telehealth applications in the literature and on the market try to fulfill a similar set of functionalities. These functionalities are identified and compared in several theoretical frameworks. However, to our knowledge, there are no practical implementations of such frameworks. An easy-to-use framework to generate customizable telehealth applications would be beneficial for health care providers, particularly for small or middle-sized providers who lack the technical knowledge or/and the budget for that. In this paper, we design and develop a framework enabling health care providers the fast development and extension of such applications.**

Keywords – telemedicine; teletherapy; framework; personal ehealth.

I. Introduction

Telemedicine and telehealth are defined as "the use of information and communication technologies (ICTs) to solve health problems, especially for people living in remote and underserviced areas" [14]. Various services can be offered by telemedicine, such as symptom assessments or the provision of information about medications [6].

The range of new telehealth applications, including changes and innovations in the digital health sector resulting from the COVID-19 pandemic, significantly increased [27]. Not only larger, mostly government or institutional providers but also private health providers in various specialization fields need to resort to telehealth solutions.

However, independently of the specialization field, these solutions share several basic functionalities [6]. For instance, the solution must enable patient management, and remote communication options.

In [31], a systematic review compares frameworks for the implementation of telehealth services. However, the focus is the contribution to the success rate of these services. Most of the selected papers (and other like [32]) discuss the evaluation of telehealth services. The framework proposed in [30] identifies 6 structural layers for the key structural components in telehealth applications along the patient journey. However, this framework (and other like [33], or [34]) is only theoretical, i.e., to our knowledge, no practical implementations are available. Therefore, the aim of this paper is to provide an easy to use and modular practical implementation for telehealth services by means of a framework.

Bearing in mind that the development of a software solution is a costly activity [17], and that telehealth applications share similar functionalities, we propose a framework which aims at the fast development of customized applications with almost no previous technical knowledge. Therefore, utilizing this framework, the health service provider can alleviate the development costs and benefit from a customized application. This framework is particularly targeted at small or middle-sized providers, who often lack a budget for software development. This framework should also fulfill several requirements among others modularity, extensibility, and scalability. Moreover, the applications generated through this framework share the same structure, leading to higher interoperability.

II. Requirements

Literature research was performed to identify the most essential functionalities of a framework for telehealth applications.

First, we identified relevant telehealth applications available on the market. We selected the following applications:

- Care01 [5]: Care01 is an application that specializes in digital surgery management. Care01 covers functions, such as appointment scheduling, patient management, but also video calls with patients.
- Clearstep [7]: Clearstep offers components for symptom checking and patient management. With the Smart Care Routing™, and the use of Artificial Intelligence (AI), the user should receive information about the health status or whether a visit to the doctor is necessary.
- Doxy.me [11]: With Doxy.me, components, such as video telephony, chats, patient management, and a dashboard are freely available. Certain functions, such as referrals can also be commercially subscribed to.

- Latido [20]: offers a product that can be integrated into existing medical software. Latido focuses on video communication and patient management. Additionally, functionalities, such as appointment scheduling or financial management are also provided.
- OpenEMR [23]: OpenEMR stands for Open Electronic Medical Record and basically offers the storage of EMRs, with which clinical data of patients, information on invoices, or medical history can be easily stored [22]. OpenEMR also offers functionalities, such as creating patient appointments or searching for medications.
- Samedi [26]: Samedi includes the following functions: online appointment scheduling, calendar and resource planning, payment function, video consultation, vaccination management, online patient forms, and a patient portal.

As a result of our literature research, we identified the following most relevant functionalities or components:

- Patient management (core): information, such as first and last names, and date of birth, should be stored.
- Video calls: enabling health service providers and their patients to communicate.
- Symptom tracking: recording symptoms of a given patient is needed to assess the course of an illness.
- Online appointments: the patient and the health service provider can book appointments with one another.
- Content management: the health service providers can create content, e.g., mini lessons, video exercises, or information to enhance health literacy, and grant access to this content to their patients.
- Interoperability & Data exportability: medical data should be interoperable so that they can be integrated in third party tools fulfilling the supported standards.
- Personalization: the application should be adaptable to the corporate design of the company. This includes defining a name, a background color, and a logo for the application.

In addition to these functionalities, we identified requirements related to the flexibility and scalability of the framework. These are:

- Open-closed principle [15]: it should be easy to add new modules/functionalities to the framework without modifying the existing code base.
- Dynamic: the user can select the modules that should be included in the application.
- Secure: data security should conform to the General Data Protection Regulations (GDPR) [12].
- Performance and availability [16]: modules and databases should be replicable to allow on-demand scalability.

III. Framework Architecture

The framework should contain the functionalities identified in Section II. An overview of the framework architecture is displayed in Figure 1.

The *patient management* functionality is the core of any application and should always be present. All other functionalities are optional and can be added to the generated application when needed.

For instance, the users can decide that their application may only contain the *video calls*, *online appointments*, and *content management* functionalities in addition to the mandatory *patient management*. This would be typical for applications where a remote support or caring, but no medical data are required: the user can book appointments which will take place through video calls, and between these appointments, the user can perform some exercises provided in the *content management*.

The implementation of this architecture applies concepts similar to the ones of microservices and modular programming [19]. Microservices should fulfill the following requirements:

- Independent databases per service.
- Independent hosting.
- Independent codebase.

As displayed in Figure 1, these requirements are only partially fulfilled by our framework. Indeed, there is only one database, and the hosting takes place on one target. This decision was made to simplify the deployment of the resulting telehealth application, since the framework should also be used by persons with little technical knowledge. However, the implementation is modular and sufficiently scalable to fulfill all requirements defined in Section II.

Figure 2 displays the database model. The database stores the data required for the functionalities explained previously and illustrated in Figure 1.

The *User* table is needed for the core functionality *patient management*. The patient data is limited to a minimum but can be extended as we will discuss in Section VI. On the one hand, we need data related to the registration for the resulting application, such as an email-address, a username, and an encrypted (hashed and salted) password.

On the other hand, we need real-life data related to the user, such as the first name, last name, and date of birth. To differentiate between patients and medical professionals, a *Role* table was added. Therefore, the user will be assigned a given role through an attribute in the *User* table.

For the *symptom tracking* functionality, we modeled the Symptom table. A symptom has a name and a description, is related to a user, and started on a given date. Moreover, we can record if the symptom is active through the eponymous attribute. This structure would be sufficient for relatively small applications. However, the splitting of the symptom name and description, as well as a code, in a separate table could be performed if the resulting application must use standards, such as SNOMED, or the International

Classification of Diseases (ICD), or Logical Observation Identifiers Names and Codes (LOINC) [4].

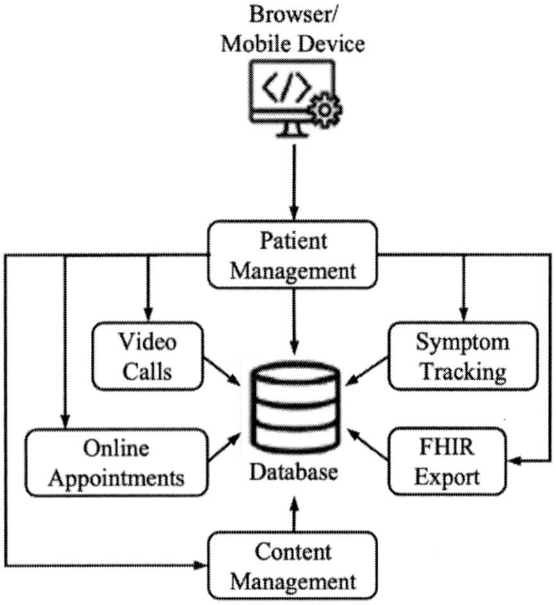

Figure 1. Framework architecture.

In order to implement the the *online appointment* functionality, the *Appointment* table records a description for this appointment, the location (for instance a link to a video call), an attribute that states if the appointment is accepted, and a reference to the two participants of the appointment. Note that the participants simply are users recorded in the *User* table. Checking if users are patients or medical professionals can be done through the role (*role_uid*) attribute of the user.

We designed two tables for the *content management* functionality: *Course* and *CourseEntry*. Thus, the content belonging to a course can modularly be organized in entries. These entries can represent textual information or exercises, in which case the *text* attribute can be used. However, they can also incorporate other media, in which case the *attachment* attribute can be utilized. The date enables the chronological ordering of the course items.

The video communication and the data export functionalities do not require their own tables. However, the data export functionality needs to access the data stored in other tables, for instance, the *User* and *Symptom* tables.

If a microservice architecture with separate databases shall be implemented, the database and the relevant tables are split. Then, a solution should be implemented to manage and grant the appropriate access to the databases connected to the various functionalities. This could be done, for instance, through Application Programming Interfaces (APIs) for each microservice.

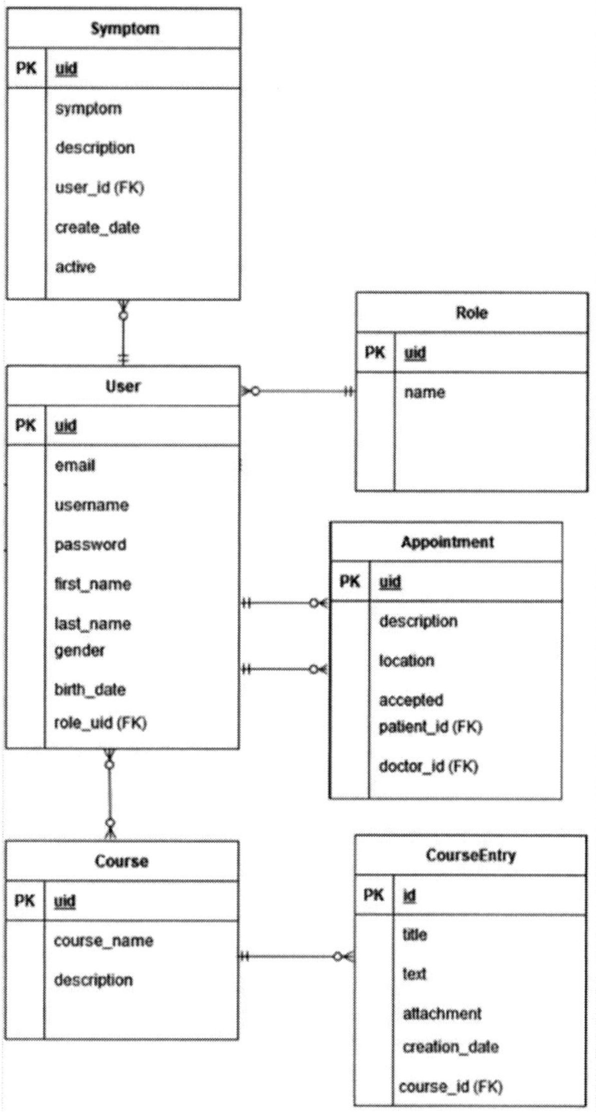

Figure 2. Database Model.

IV. TECHNICAL DESCRIPTION OF THE FRAMEWORK DEVELOPMENT

The implementation of this framework involved several tools.

MariaDB [18] was used as the database. The database was installed locally for development. It is a relational database and freely available. It has a similar implementation to MySQL but is a fork of the MySQL project. We chose a relational database for its ACID (atomicity, consistency, isolation, and durability) compliance and its transaction capabilities.

The programming language **Java** [28] in its version 11 was chosen for the backend and the initializer (see Section VI). Java was preferred due to its platform independence, its relative popularity among developers over time, and the availability of further frameworks such as the Spring Framework.

The **Spring Framework** offers several functionalities in the form of standalone projects for web applications [8]. The following Spring projects were used as part of the development:

- Spring Boot: for a standalone application that can replace or embed resources, such as a web server.
- Spring Web: for the simplified creation of web Model View Controller (MVC) applications.
- Spring Java Persistence API (JPA): as a persistence framework alternative to Hibernate for database operations.
- Spring Security: for user administration and security in the application, as well as for authentication, authorization, and registration.
- Spring Session: for managing the sessions of the logged-in users.
- Spring Websocket: for the creation of web sockets, needed for instance in the video communications.

WebRTC [3] is a technology or standard that offers free video communication.

In the frontend, **HTML** was used as the basis, **JS** for client-side operations and **CSS** for styling [25].

Maven was used as the build and configuration tool [9]. Gradle can be used as an alternative.

V. TECHNICAL DESCRIPTION OF DEPLOYMENT

As we explain in Section VI, the framework allows two use cases: initializer and application development kit.

To perform the former use case, the initializer can be hosted on a public website or locally.

For the latter use case, the framework components are simply checked out and the required configurations, such as the connection information (IP, etc.) of the database must be completed.

Since all components are Spring Boot applications, no separate web server, such as Tomcat and Glassfish is required. However, Spring Boot still offers the option of using a different web server.

The components are all implemented as independent web applications. This means that they must also be accessible on different ports. Six ports must therefore be reserved for the six components. During development, an additional port must also be reserved for the initializer. These seven ports are set per default (8080-8085 and 8090) and can be configured in the property files, if needed. The port configuration and protection should be carefully performed to minimize security risks and having several open ports might imply more overhead for this task.

Each component and the initializer have their own property file, which can be accessed from outside the compiled component. This implies that the port and other settings can be adjusted without having to rebuild the component and create a new resource file. On the one hand, similarly to port configuration, having many property files could increase the complexity and potentially the security risks. On the other hand, a centralized property file would be a single point of failure and would be less flexible.

If this number of ports is too high or unsuitable for practical use, a reverse proxy can be used. This maps requests and stands between the clients and the internal server, i.e., requests to the one public port are mapped to internal ports. Therefore, only one public port would be needed and declared as an open port in the reverse proxy configuration. Nginx offers a possible implementation of reverse proxies [2]. Obviously, the use of a reverse proxy would add another layer of complexity and potential points of failure to the system.

Docker can also be used to create the infrastructure more easily. With Docker, the application and its dependencies can be encapsulated [24]. The Docker image and container can then be deployed independently of the underlying operating system and enables more scalability. Nginx can also be used within Docker to implement a reverse proxy and thus minimizes the opened public ports as mentioned earlier.

VI. USE CASES

The framework can be used mainly in two ways:

- Initializer or application generator: the users just want a way to produce a telehealth application in a few clicks by selecting the functionalities they desire and optionally their corporate design.
- Application development kit: the users modify and enhance the framework with their own functionalities and customizing. The framework is then like a Software Development Kit (SDK), i.e., the core of the application is already available.

When utilizing the initializer, the user will go through several steps leading to the generation of the desired application. The user is guided through the process with help of descriptions and explanations in each step.

In the first step or graphical user interface (GUI), the user can personalize the appearance of the resulting application. This is the chosen way to fulfill the personalization functionality identified in Section II. Indeed, in the initializer, the users can personalize the application in three ways: a name corresponding to the name of their company or product, a color corresponding to the corporate design of their company, and a picture corresponding to the logo of the company or product.

In the second step, the users select, by means of checkboxes, the components, or functionalities that they want to include in the generated application. The available components are the ones listed in Section II: *video calls, symptom tracking, online appointments, content management, interoperability & data exportability.* Note that there is no checkbox for the patient management component since this component is mandatory and therefore, always part of the generated application.

Then, the user can click on a *Download* button and get a zip file containing all the resources needed to start the telehealth application. The resources are stored in three directories: a *config* directory containing the configuration files for each of the selected components, a *file* directory containing the database creation scripts, and a *module* directory containing the generated components as individual jar-files. The database still needs to be created using the included scripts. Subsequently, the user can already start the application.

When utilizing the framework as an SDK, the developer only needs to download the sources and develop the desired extensions or modifications.

VII. LIMITATIONS

Economical, ethical, social, and legal aspects might not have been considered to their full extent. The aim of this paper was to present a prototype, not a market-ready product.

Telemedicine and telehealth result in cost savings through reduced expenses for examinations in the clinical facilities or travel times [29], as patients also receive treatment at home, and could decrease the waiting times or bed occupancy in hospitals [1].

The framework proposed in this paper aims at reducing the effort required to design and develop a new software architecture for telehealth applications, even if some tasks for customization and deployment remain. However, new or existing expenses should be taken into account, such as investments in new hardware and software, organizational and structural integration into existing supply structures [10].

Several ethical aspects should be considered when using telemedicine and telehealth, such as the difference in quality of the collected data. The quality of the data can have an impact on the quality of the treatment and the trust between medical staff and patients [1].

Patients may also be reticent to agree to a continuous tracking and monitoring, even if this monitoring could lead to a more successful treatment [21].

Telemedicine and telehealth can be a means against the shortage of doctors in rural areas and poorer care in remote areas, as patients can communicate and have access to health services remotely.

However, it must also be noted that this might lead to further concentration of the medical infrastructure in bigger cities.

Moreover, the personal contact between patients and health service providers, as well as between patients and their relatives, for example when accompanying them to doctor's visits could be gradually reduced [21]. Since social contacts contribute to the health of patients, this aspect would be counter-productive.

In this paper, we developed a framework for telehealth and took legal aspects, such as GDPR into account. The use of Spring Security enables the fulfillment of several requirements of the GDPR through authentication, authorization, and user administration and security in the application. However, the users are responsible for the security of the database – for instance the strength of the password for the root users – which could be a security weakness.

Other questions regarding the liability, attribution, distribution, copyright, and warranty of the framework are still open. For instance, the following options are available for offering the framework: open source, GNU, or Creative Common (CC).

Another question is to which extent the framework or part of the framework fall under the Medical Device Regulations (MDR) [13].

VIII. CONCLUSION AND FUTURE WORKS

In this paper, we propose a framework for the fast development of customized telehealth applications based on requirements identified in the literature and in applications available on the market (see Section II). We present an architecture fulfilling these requirements (see Section III), show how to implement it (see Section IV) and deploy the product (see Section V). After that, we explain how the user can create an application with help of this framework or extend its functionalities (see Section VI). Finally, we identify several limitations for this framework (see Section VII).

As described in this paper, we tried to focus on the most wide-spread functionalities found in telehealth applications. The implementation of further components or functionalities is obviously a topic of interest for future implementations. For instance, one could evaluate the interest of an AI-chatbot, or components dedicated to accounting tasks.

We also mentioned that the recorded data are only partially following standards in the framework. To enable an easier integration with other tools, more data interoperability, reached through the implementations of standards, would be an advantage.

We are aware that scalability is a challenge, and we show how Docker can answer this question. However, with an increasing number of components, Kubernetes could be a better choice for the management of the containers.

Finally, we would like to evaluate the most appropriate model (open source, GNU, CC, proprietary) and make this framework available to the public.

ACKNOWLEDGMENT

We thank our colleague Anita Töchterle for the comments and proof-reading that greatly improved the manuscript.

REFERENCES

[1] A. Atac, E. Kurt, and S. E. Yurdakul, "An overview to ethical problems in telemedicine technology", Procedia-Social and Behavioral Sciences, vol.103, pp. 116–121, 2013.

[2] A. Baeldung, "Serving Multiple Proxy Endpoints Under a Location in Nginx", 2022.

https://www.baeldung.com/linux/nginx-multiple-proxy-endpoints. Retrieved: February, 2024.

[3] N. Blum, S. Lachapelle, and H. Alvestrand, "Webrtc: Real-time communication for the open web platform", Communications of the ACM, vol. 64, no. 8, pp. 50–54, 2021.

[4] O. Bodenreider, R. Cornet, and D. J. Vreeman, "Recent developments in clinical terminologies — snomed ct, loinc, and rxnorm", yearbook of medical informatics, vol. 27, no. 01, pp. 129–139, 2018.

[5] Care01. https://www.care01.com/. Retrieved: February, 2024.

[6] M. F. Chiang, J. B. Starren, and G. Demiris, "Telemedicine and Telehealth", Biomedical Informatics, pp. 667-692, 2021, doi:10.1007/978-3-030-58721-5_20.

[7] Clearstep. https://www.clearstep.health/. Retrieved: February, 2024.

[8] I. Cosmina, R. Harrop, C. Schaefer, and C. Ho, "Pro Spring 6: An In-Depth Guide to the Spring Framework", Apress, July 2023. ISBN: 9781484286401.

[9] M. Tyson, "What is Apache Maven? Build and dependency management for Java.", 2020. https://www.infoworld.com/article/3516426/what-is-apache-maven-build-and-dependency-management-for-java.html. Retrieved: February, 2024.

[10] S. Demirci, F. Kauffeld-Monz, and S. Schaat, "Perspectives for Telemedicine – Prerequisites of the Scalability and Market Potential" (original title: "Perspektiven für die Telemedizin – Voraussetzungen der Skalierung und Marktpotenzial"), IIT Berlin, May 2021.

[11] Doxy.me. https://doxy.me/en/. Retrieved: February, 2024.

[12] European Commission, "Regulation (EU) 2016/679 of the European Parliament and of the Council of 27 April 2016 on the protection of natural persons with regard to the processing of personal data and on the free movement of such data, and repealing Directive 95/46/EC (General Data Protection Regulation) (Text with EEA relevance)", European Commission, 2016.

[13] European Parliament, "Regulation (EU) 2017/745 of the European Parliament and of the Council of 5 April 2017 on medical devices, amending Directive 2001/83/EC, Regulation (EC) No 178/2002 and Regulation (EC) No 1223/2009 and repealing Council Directives 90/385/EEC and 93/42/EEC (Text with EEA relevance.)", European Parliament, 2017.

[14] "Fundamentals of Telemedicine and Telehealth", edited by Shashi Gogia, Academic Press, Elsevier, 2019. ISBN: 978-0-12-814309-4.

[15] E. Gamma, R. Helm, R. Johnson, and J. M. Vlissides, "Design Patterns: Elements of Reusable Object-Oriented Software", Addison-Wesley Professional, 1994, ISBN: 0201633612.

[16] R. Jiménez-Peris and M. Patiño-Martínez, "Replication for Scalability", in: L. Liu and M. T. Özsu (eds), Encyclopedia of Database Systems, Springer, Boston, MA, 2009. doi:10.1007/978-0-387-39940-9_314.

[17] M. Jorgensen and M. Shepperd, "A Systematic Review of Software Development Cost Estimation Studies", in *IEEE Transactions on Software Engineering*, vol. 33, no. 1, pp. 33-53, Jan. 2007. doi: 10.1109/TSE.2007.256943.

[18] E. Kenler and F. Razzoli, "MariaDB Essentials", Packt Publishing Ltd, 2015.

[19] M. Larsson, "Microservices with Spring Boot 3 and Spring Cloud", 3rd edition, Packt Publishing Ltd, Aug. 2023. ISBN: 9781805128694.

[20] Latido. https://latido.at. Retrieved: February, 2024.

[21] A. Manzeschke, "Telemedizin und ambient assisted living aus ethischer perspektive", Bayrisches Artzeblatt, 2014.

[22] J. Noll, S. Beecham, and D. Seichter, "A qualitative study of open source software development: The open emr project", in 2011 International Symposium on Empirical Software Engineering and Measurement, pp. 30–39. IEEE, 2011.

[23] OpenEMR. https://www.open-emr.org/. Retrieved: February, 2024.

[24] N. Poulton, "Docker Deep Dive", Nielson Book Services, Mai 2023. ISBN: 978-1916585256.

[25] J. N. Robbins, "Learning Web Design - A Beginner's Guide to HTML, CSS, JavaScript, and Web Graphics", 4th edition, O'Reilly, 2012. ISBN: 978-1-449-31927-4.

[26] Samedi. https://www.samedi.com/en/. Retrieved: February, 2024.

[27] J. Shaver, "The State of Telehealth Before and After the COVID-19 Pandemic", Prim Care, vol. 49, no. 4, pp. 517-530, 2022. doi: 10.1016/j.pop.2022.04.002.

[28] K. Sierra and B. Bates, "Head First Java", 2nd edition, O'Reilly Media, Inc., 2005. ISBN: 0596009208.

[29] C. L. Snoswell et al., "Determining if telehealth can reduce health system costs: scoping review", Journal of medical Internet research, vol. 22, no. 10, 2020. doi: 10.2196/17298.

[30] A. Prakash et al., "Building A Global Framework For Telehealth", Health Affairs Forefront, June 27, 2023. doi: 10.1377/forefront.20230621.134595

[31] L. A. van Dyk, "A review of telehealth service implementation frameworks", Int J Environ Res Public Health, vol. 11, no. 2, pp. 1279–1298, 2014. doi: 10.3390/ijerph110201279.

[32] "Creating a Framework to Support Measure Development for Telehealth", National Quality Forum, August 2017.

[33] D.W. Ford, J. Harvey, J.T. McElligott, and S. Valenta, "TSIM: The Telehealth Framework - A comprehensive guide to telehealth implementation and optimization", Stationery Office Books, 2021. ISBN: 9780117092969.

[34] "ACRRM Framework and Guidelines for Telehealth Services", Australian College of Rural and Remote Medicine, June 2020.

[35] TIOBE Index. https://www.tiobe.com/tiobe-index/ Retrieved: February, 2024.

A Temporal Perspective on Electronic Medicine Management Work

Line Lundvoll Warth
UiT—The Arctic University of Norway
Norwegian Centre for E-health Research
Tromsø, Norway
Email: line.lundvoll.warth@uit.no

Kari Dyb
Norwegian Centre for E-health Research
UiT—The Arctic University of Norway
Tromsø, Norway
Email: kari.dyb@ehealthresearch.no

Abstract—**Poor communication between professionals across levels of care regarding patients' medication can lead to errors. Norway has invested in several electronic initiatives to improve collaboration in medicine management. This study elaborates on collaborative electronic medicine management work in specialist and primary care, asking what the problems with electronically shared medicine information are and how they can be solved. Based on *community of practice* as a method, three focus group discussions were arranged with healthcare professionals in 2022 and 2023. One theme was selected: the mismatch between the medication list in the discharge summary and the medication list before hospitalization. The analysis illustrates that this mismatch is connected to temporality in the patient's illness trajectory, their medicines, and the professionals' work related to this temporality. Overall, this paper contributes to sociotechnical perspectives on eHealth practices, focusing on structures and collaborative work processes. New eHealth initiatives offering digital tools for medicine management must take the temporal structures of medicine management work into account.**

Keywords-electronic medicine management; medicine inconsistencies; information sharing; temporality; community of practice.

I. INTRODUCTION

Poor communication regarding patients' medication across healthcare levels may lead to potentially harmful medication errors [1]. When hospitalizing a patient, information about medicines is shared by the General Practitioner (GP) in primary care with specialist care and, later, vice versa. In Norway, governmental strategies encourage digital communication and collaboration between levels of care to make patient health information available for professionals across levels throughout the patient's trajectory [2]. Medicine management and digitalization are high priorities in the Norwegian e-health strategies [3]. To improve medicine management, several national plans and electronic initiatives have been invested in [4], including electronic prescriptions (2004–2005) [5], the Summary Care Record (2008–2009) [2], eMultidose (2014–2015) [6], and the Patient Medications List (2017–2018) [7]. These electronic initiatives for medicines management may be independently more or less successful, but they have been unsuccessful in coordinating all the information and managing all the work involved.

The existence of multiple electronic information systems poses a challenge for medicine management, as none of them provides a complete overview of a patient's medicine history. The barriers to exchanging health information are connected to incomplete information and inefficient workflows that do not meet the needs of professionals [8]. This is not a problem unique to Norway. Denmark, Finland, Sweden, and Norway are all at different stages of implementing national shared medication systems to increase access to information and reduce discrepancies between medical information. Despite the implementation of new systems, the information in the lists is not always up to date, systems are not integrated, there is a gap between the integration of information and appropriate working routines, and there are legal and technical issues [9].

To reduce the number of medication errors and improve the quality of care, healthcare professionals need to collaborate regarding medicines and treatment. When a patient is referred between primary and specialist care, the professional's main digital communication tools are the referral letter from primary care to the hospital and the discharge summary from the hospital to primary care. Information about the patient's medication is outlined in the medication list. Collaboration and continuity of care in terms of adherence to medication lists when patients are transferred from one health care level to another are challenging [10]. A study of patients' medication lists documented during their hospital admissions, hospital stays, and return to their GPs reports that half of the errors found originated from incomplete medication lists provided in referral letters [1]. Another study highlights the problem of changes made in medication during hospital stays that are not always explained in discharge letters, making it difficult for GPs to follow up on the medication as intended [11]. Hence, there are inconsistencies between patients' prescribed medication on admission to hospital and their prescribed medication upon discharge [12].

Several digital systems contribute to the work process related to information sharing. This study does not elaborate on one special technology or system; instead, it focuses on the exchange of medicine information in electronic systems per se, specifically those used for medicine management. The transition between levels of care in an illness trajectory is particularly challenging because of the work related to information exchange. Using physicians' and GPs' points of views, this study elucidates the work related to the sharing of information regarding patients' medicines between the hospital and the primary care institution. The study asks the following questions: *What are the problems with electronically shared information and how are they solved?* This paper contributes to the literature on electronic medicine management and the body of empirical and analytic investigations of temporality in collaborative work and medicine management.

The rest of this paper is organized as follows: Section II describes the theoretical framework, Section III elucidates the method used, Section IV presents the results and analysis, Section V outlines the discussion, and Section VI provides the conclusions and suggestions for future work.

II. Theoretical Framework

We frame the work related to electronic medicine management from a temporal perspective. According to Orlikowski and Yates [13], studies of temporality are mainly rooted in one of these two perspectives: time is understood as subjective, existing independent of human action, clock-based, and measurable, or it is viewed objectively as a phenomenon socially constructed through human action. This subjective–objective dichotomy is often presented as the contrast between clock time and event time [14]. Orlikowski and Yates [13] propose joining the subjective and objective into a practice-based perspective in which time is both independent of and dependent on human actions. According to this practice-based perspective, temporality is explored in terms of people working and interacting with the temporal features of their work [13].

Furthermore, temporality can be regarded as the sequence of work activities that occur as the patient progresses through a particular unfolding illness trajectory [15]. While exploring the work related to a particular patient as their illness unfolds, the work of managing that illness also unfolds. For example, the patient's illness goes through various stages, demanding their transfer between healthcare levels and involving different professionals who make changes to their medicines.

Utilizing Orlikowski and Yates [13], this study combines the subjective and objective perspectives to analyze temporality from *a practice-based perspective*. The study uses the *medicine management trajectory* to illustrate the *timeline* and *work activities* performed. Electronic information systems (e.g., medicine lists) are the tools that make medicine information sharing possible. Based on these perspectives, the problems and solutions related to GPs' handling of medication inconsistencies are explored in this study.

III. Method

This is a qualitative study exploring the practices and dynamics of a group of professionals. It uses a Community of Practice (CoP) approach to emphasize the social nature of learning and the importance of shared practice, mutual engagement, and a shared domain of interest [16]. The CoP serves as an arena for healthcare professionals to meet and share knowledge, expertise, and best practices related to their work with electronic medicine management. Professionals were recruited who shared a common interest in renewing their medicine practices, engaging in joint activities, and developing a shared repertoire of resources and knowledge.

In Norway, two new hospitals are under construction. These hospitals were selected for this study because professionals in these hospitals were eager to discuss challenging cases, share successful strategies, and learn from each other's expertise. These professionals were established as the core group and replaced by new participants if participants from the core group were unable to participate. The main aim was to gather a group of interprofessional participants who represented the whole chain of medicine management and were willing to share their expertise. The empirical material was collected from three in-person focus group discussions [17]. Two of these discussions were held in 2022 and one in 2023 (Table 1). The discussions will continue as the project progresses. Furthermore, eight to ten participants were part of each discussion. The participants were encouraged to present 1) a case that each of them considered challenging in terms of medicines management and interaction between levels and 2) the importance of national guidelines for local medicines management practices. The case in this paper is chosen because the participants considered it as a classical example, engaging them all in a shared engagement with and interest in the problem and a mutual interest in finding a solution.

The focus group discussions lasted approximately three hours each, during which participants were presented with the challenges of electronic medicine management. The discussions were audio-recorded and transcribed. The data were analyzed using thematic analysis, identifying, analyzing, and reporting themes in the empirical data [17]. For this paper, the findings focusing on hospital physicians' and GPs' work with medicine management and their reflections on problem-solving were chosen. Therefore, the GPs' and physicians' utterances were selected. Nevertheless, the meaning of the content was produced by an interprofessional group dynamic. The results reflect the patterns that emerged from the findings arising from the three discussions.

TABLE I. PARTICIPANTS

Year	Represented Professionals in the CoP
2022	Nursing home doctors (2), hospital physician, hospital nurse, nursing home nurse, home care nurse, representative information and communication technology (ICT), community pharmacist, and hospital pharmacist
2022	Nursing home doctor, hospital nurses (4), nursing home nurse, home care nurse, ICT representative, hospital pharmacist, and community pharmacist
2023	GP, hospital nurse, home care nurse, hospital pharmacists (2), and community pharmacists (2)

IV. Results and Analysis

The results illustrate how medicine inconsistencies were handled in terms of (a) evaluating the problem, (b) evaluating the consequences, and (c) making adjustments to solve problems caused by temporality in electronic medicine management work.

A. *Problems with the Information Shared*

The hospital physician described a classic example of electronic medicine management work performed when a patient is admitted to and later discharged from the hospital:

> A classic example: There are several changes in the medicines [at the hospital]. The physician writes a note (...). In addition, he copies and pastes text from the previous medicines. The GP gets it back [discharge summary]. If there is a medication list that does not match what the GP remembers the patient was on before hospitalization, the GP is in a dilemma: does this mean that the hospital has decided the changes that have been made should be reversed and that he [the patient] should go back to what he used before hospitalization, or does it mean that it is a copied and pasted [version], showing that it is an error? What exactly is the hospital trying to communicate?
>
> (Physician in hospital)

In this excerpt, the physician describes the various stages of working with the information in the patient's medication list. The dilemma is regarding the mismatch between the medicine list in the referral letter and the medicine list in the discharge summary. This problem description of the dilemma caused by information sharing in the treatment trajectory was confirmed by the GP in primary care. From the perspective of time, the sequence of work activities as the patient goes through the treatment trajectory creates an expectation of medicine reconciliation. When information does not match the GP's situated knowledge of the patient, they face a dilemma. Are there old follow-up errors (e.g., copied and pasted medicine information), or has the physician in the hospital made an active reconciliation of the medicines?

From the perspective of time, this problem description is about the temporal context of medicine information. Time is constructed here through the discharge summary as an end point for the physicians' work in the hospital. The information becomes static, and the medicine list is not open to negotiation. When the patient approaches the GP, the GP sees inconsistencies in the medicine information due to which the patient's treatment is organized in returning events. As GPs often look after the same patients over a long life span, they have historic practice-based knowledge of their patients' medicine histories. As a result, temporality is represented in the information, while the information from the hospital has a here-and-now point of departure.

B. *Evaluating the Consequences of Inconsistencies*

Evaluating the consequences of information is part of the treatment practice. This assesses the severity of errors. The GP described this issue as follows:

> Many of the errors reported are errors that go well, nevertheless. There are many errors that do not have consequences (...). Most errors are not critical. If you don't get the dosage you should have for a day or two, it's mostly okay. What is not okay are the occasions they are serious.
>
> (GP)

The GP described how the consequences of information inconsistencies are evaluated according to time and effects. The wrong dosage over a short period of time was described as a non-dangerous consequence of an error. Here, the subjective temporality of clock time is connected to the evaluation of the objective, which is the temporality of the event. The temporality of the event provides the opportunity to change the dosage after a second, practice-based evaluation of the illness trajectory.

Furthermore, the GP provided an example of a non-dangerous error in the discharge summary. This error involved dietary supplements and vitamins: "You choose fights that are important. If they [the physicians at the hospital] forget a dietary supplement [in the list], they [the patient] won't die from it."

Here, the GP evaluated the consequences of errors, considering the effects of prescribing wrong medication and the effort put into the extra workload. In the quotation, "fights that are important" points to the extra work of checking information with the hospital and evaluating the consequences of the medicines taken. Information from the hospital is viewed as temporal in practice. In the same discussion, a pharmacist evaluated errors from another point of view, saying, "Yes, they should remember to take it [the dosage of dietary supplements or vitamins]!" The pharmacists wished to close the temporal event, reducing temporality, while the GP evaluated the *big picture* as a repeated element, depending on time and based on the relative importance of one medicine compared to other medicines.

The GP evaluated the consequences of information inconsistencies in relation to the severity of incorrect or missing medication:

> However, if something goes wrong, it will take a lot for it to be a major disaster. It's good to get it in [the medical lists] and communicate it accurately. Once they get a dietary supplement, they [the patients] are happy to continue. It's not like they change dietary supplements.
>
> (GP)

This excerpt illustrates how the GP not only evaluates the consequences of a potential error but also reflects on how long-term dietary supplements or vitamins are prescribed more stably over time. From the perspective of time, this evaluation reflects the temporality during the patient's treatment trajectory. Hence, the temporality in medicine is also connected to *which* medicines are taken during the patient's trajectory.

C. Solving the Problem of Inconsistencies

In the CoP, the professionals discussed opportunities to prepare for best practice. When discussing how the problem of medical information inconsistencies between the hospital and the GP could be solved, the GP suggested the following:

> Can you [the hospital] write whether you have reconciled the medicines? Is it a copy–paste [job]? Can you express whether something has been done and thought about? It's almost impossible to determine where to start because you don't really know what has been decided (...) in the end (...)—what reflection is given. The physician who takes over [in the hospital] needs to communicate so that adjustments can be made.
>
> (GP)

Solving the problem of inconsistencies can be performed by providing information on whether the medicines have been reconciled. The GP demands a decision or reflection on the information given. These assessments are the foundation of the new assessments conducted by the GP. Within the hospital, the illness trajectory represents temporality. The patient moves between departments, and information needs to be communicated to obtain a complete overview when the patient leaves the hospital. This information is also the foundation of the adjustments made. In this quotation, the GP illustrates how their work involves individually preparing temporality through making medical adjustments and planning the future.

If nurses reveal inconsistencies in the information that may lead to errors, they circulate information about the action, reattempting the decision-making of the physician or the GP by using the telephone. As the home care nurse said, "Well, then we will call." To this, the hospital physician responded as follows: "No! If I'm unsure, I do what I think is right. I am the one who decides (...). It is not the others' task to make assessments. It is a physician's task."

The home care nurse perceived the information as an open-ended discussion, attempting to collectively create an emergent temporal structure by searching for knowledge. The physician viewed this as an individual responsibility, considering that in the end they have to close the negotiation by evaluating and deciding medicines and further treatment. Here, temporality is represented in the information, which the GP or the physician needs to stabilize until new information occurs. This illustrates the temporal future of the work: that is, temporality as practice-based.

V. Discussion

The results illustrate how the problems associated with electronically shared information are related to temporality. The temporality is both subjective, as the patient goes through a treatment trajectory that is clock-based and connected to time and place, and objective, connected to events that are socially constructed by the patient's and professionals' actions. Temporality is made visible in the patient's illness trajectory that changes, in the medicines that change, and in the professionals' work that requires adjustments. The problem with electronical shared information is illustrated by the medicine list in the discharge summary. In this paper, this is revealed by the GP who finds information that does not match the GP's situated knowledge of the patient. Practically, it is shown by examples of old follow-up errors (i.e., copied, and pasted medicine names) and active reflections on the medicine list.

Previous studies have highlighted, among other issues, incomplete medication lists in referral letters meant for the hospital [1] and insufficient information in discharge letters, which make it difficult for the GP to follow up [11]. Instead of evaluating the medicine list itself or attributing errors to professionals, this study uses the theoretical framework of temporality to shed light on the properties of practice, time, and events in professional medicine management processes. The work related to medication at the hospital is an event that occurs during the hospital stay, and the timeline is completed when the patient leaves the hospital. Hence, the medicine list has an endpoint when the patient is transferred from the hospital to their home, a nursing home, or home care, becoming the responsibility of the GP. The hospital performs its work related to medicine in a more closed and deadline-oriented manner. By contrast, the temporal structure of primary care is more open-ended and event-based.

Both the physician in the hospital and the GP understand the illness trajectory as temporal and in progress. They have a circular approach to patient illnesses and constantly wish to know what has been happening (looking backward) and what is planned (looking forward) for the patient, attempting to shape the treatment trajectory. They continuously try to find past information and consider what future information they will need [15]. This temporal future of the work (i.e., practice-based temporality) [13], which is connected to patients' medicines and movements in the healthcare system, is the professional and organizational working-life structure. Overall, the trajectory *in house* is temporal because physicians or GPs are responsible for the treatment of patients from the minute they arrive to the moment they leave. Nevertheless, at the hospital, the work related to electronic medicine management is an event representing the closure of work when the patient is discharged from the hospital to a primary care institution. Here, medicine management has a temporal structure that professionals currently use in their everyday work.

When the patient is the responsibility of the GP and potential errors exist in the medicine list, the GP evaluates how to solve this issue in terms of the degree of its consequences over time. By contrast, the nurses and pharmacists evaluate this as an open-ended discussion, attempting to collectively create an emergent temporal structure. The GP has responsibility for the patient and makes decisions on the basis of the patient's history and current situation [15]. A previous study of the Summary Care Record in Norway [18] shows that doctors did not trust manually updated information in the system. In this paper, the findings illustrate how the GP evaluates information, trusting their own judgment and situated knowledge of the patient. Hence, the study supports previous findings, showing that the GP conducts assessments and makes decisions based on current and previous knowledge of patients' medicines. The changes made by the GP contribute to a constant movement in medicine management, in which assessments and trust are linked to the physician's or GP's individual competence.

Moreover, electronic tools offer opportunities to manage medicine information and produce and negotiate the temporal order of professionals. The tools, which are the referral letter from primary care to the specialist service and the discharge summary from the hospital to primary care, serve as elements initiating discussions of gaps. To elaborate, professionals discuss and search for proposals and agreements about treatment, leading to additional information gathering about medicines. Hence, the patients' medicines constitute a form of temporality in practice.

Developing technology for medicine management must take temporal structures into account. Furthermore, complex medicine management is interdisciplinary work. Work attempting to solve inconsistencies has a different character among different professionals (i.e., different perspectives regarding the degree of medication errors). The discharge summary from the hospital is situationally dependent, representing the here and now, while the GP's practice has a lifetime perspective. Moreover, problem-solving, such as that conducted by nurses and pharmacists, is performed with different strategies by different professionals. For example, nurses call for checkups, working with the medicine list with a clear end in mind. Pharmacists account for all medicines, including vitamins, at this temporal endpoint. Hence, the medicine list accounts for the different temporalities among various professionals. New investments and the implementation of new technologies must take this temporality into account.

VI. CONCLUSIONS AND FUTURE WORK

Medicine management work is complex. The temporal organization of work is considered a practical accomplishment of human activities. The results illustrate that the problems associated with electronically shared information are related to temporality in the patient's illness trajectory, their medicines, and the work of professionals related to this temporality. Furthermore, electronic management systems are stable, while the illness trajectory, medicines, and work of professionals are only stabilized for a short period of time. On the one hand, at the hospital, medicine management regarding one specific patient is event-based, with a beginning and an end to the diagnosis, where the discharge summary represents the closure of an event. On the other hand, in primary care, the treatment practice is more temporally structured, whereas the GP uses a life-course perspective in the treatment of the patient, creating a temporal structure whose character is derived from aspects of the working-life structure. To solve this complexity, new initiatives involving digital tools for medicine management need to take into account the temporal structure of future work (i.e., practice-based temporality) and connect it with a tool that facilitates temporal medicine information through the healthcare trajectory. The temporality is neither subjective nor objective; rather, it involves a coordination between the different temporalities. This study is concerned with the collaborative work involved in exchanging medical information per se. A limitation of this study is regarding its practical adaption to the development of technology. Future research should continue to explore medicines management work practices to provide further knowledge about eHealth systems and how they can take into account the complexity of interprofessional medicines management work.

ACKNOWLEDGMENTS

We thank the Norwegian Research Council for funding this project: Electronic Medicine Management (eMM)—A Comparative Case Study Promoting Coherent Health and Welfare Services (NFR314382). We thank all the healthcare professionals who participated in the study.

REFERENCES

[1] K. Frydenberg and M. Brekke, "Poor communication on patients' medication across health care levels leads to potentially harmful medication errors," *Scand. J. Prim. Health Care*, vol. 30, pp. 234–240, Dec. 2012, doi:10.3109/02813432.2012.712021.

[2] Report No. 47 to the Storting (2008–2009). *The Coordination Reform: Proper Treatment—At the Right Place and Right Time.* [Online]. Available from: https://www.regjeringen.no/en/dokumenter/report.no.-47-to-the-storting-2008-2009/id567201/ [retrieved: April, 2024].

[3] *National eHealth Action Plan 2017–2022.* [Online]. Available from: https://www.ehelse.no/publikasjoner/nasjonal-handlingsplan-for-e-helse-2017-2022/ [retrieved: April, 2024].

[4] Report No. 9 to the Storting (2012–2013). *One Citizen—One Journal.* [Online]. Available from: https://www.regjeringen.no/no/dokumenter/meld-st-9-20122013/id708609/ [retrieved: April, 2024].

[5] Report No. 18 to the Storting (2004–2005). [Online]. Available from: https://www.regjeringen.no/en/dokumenter/report-no.-18-to-the-storting-2004-2005/id406517/ [retrieved: April, 2024].

[6] Report No. 28 to the Storting (2014–2015). *The Drug Notification: Correct Use—Better Health.* [Online]. Available from: https://www.regjeringen.no/no/dokumenter/meld.-st.-28-20142015/id2412810/ [retrieved: April, 2024].

[7] Meld. St. 15 (2017–2018). *A Full Life—All Your Life: A Quality Reform for Older Persons.* [Online]. Available from: https://www.regjeringen.no/en/ocumenter/meld.-st.-15-20172018/id2599850/ [retrieved: April, 2024].

[8] K. B. Eden et al., "Barriers and facilitators to exchanging health information: A systematic review," Int. J. Med. Inform., vol. 88, pp. 44–51, April 2016, doi.org/10.1016/j.ijmedinf.2016.01.004.

[9] T. Hammar et al., "Nationally shared medication lists: Describing systems in the Nordic countries." Stud. Health Technol. Inform., vol. 302, pp. 207–211, 2023, doi.org/10.3233/SHTI230104.

[10] N. Van Sluisveld, M. Zegers, S. Natsch, and H. Wollersheim, "Medication reconciliation at hospital admission and discharge: Insufficient knowledge, unclear task reallocation and lack of collaboration as major barriers to medication

safety," BMC Health Serv. Res., vol. 21, 170, June 2012, https://bmchealthservres.biomedcentral.com/articles/10.1186/1472-6963-12-170.

[11] A. Bergkvist et al., "Improved quality in the hospital discharge summary reduces medication errors—LIMM: Landskrona Integrated Medicines Management," Eur. J. Clin. Pharmacol., vol. 65, pp. 1037–1046, 2009.

[12] B. Glintborg, S. E. Andersen, and K. Dalhoff, "Insufficient communication about medication: Use at the interface between hospital and primary care," Qual. Saf. Health Care, vol. 16, pp. 34–39, 2007.

[13] W. J. Orlikowski and J. Yates, "It's about time: Temporal structuring in organizations," Org. Sci., vol. 13, pp. 684–700, Dec. 2002, doi.org/10.1287/orsc.13.6.684.501.

[14] E. Jacques, The Form of Time. London: Heinemann, 1982.

[15] A. Strauss, S. Fagerhaugh, B. Suczek, and C. Wienner, Social Organization of Medical Work. Chicago: University of Chicago, 1985.

[16] E. Wenger, Communities of Practice: Learning, Meaning, and Identity. Cambridge: Cambridge University Press, 1998.

[17] V. Braun and V. Clarke, "Using thematic analysis in psychology," Qual. Res. Psychol., vol. 3, pp. 77–101, July 2006, doi.org/10.1191/1478088706qp063oa.

[18] K. Dyb and L. L. Warth, "The Norwegian National Summary Care Record: A qualitative analysis of doctors' use of and trust in shared patient information," BMC Health Serv. Res., vol. 18, 252, April 2018, doi:10.1186/s12913-018-3069-y.

Load Induction then Simultaneous Relaxation: Insights from Multi-Modal Time-Series Data Measured with Low-Cost Wearable Sensors

Christoph Anders, Sai Siddhant Gadamsetti, Nico Steckhan, Bert Arnrich
Digital Health - Connected Healthcare
University of Potsdam
Hasso Plattner Institute
14482 Potsdam, Germany
e-mail: Christoph.Anders@hpi.de, Siddhant.Gadamsetti@hpi.de, Nico.Steckhan@hpi.de, Bert.Arnrich@hpi.de

Abstract—**Prolonged levels of high mental workload and resulting stress are among the main causes of employee sickness. A possible solution would be implementing business rules based on objective analyses of stress levels and cognitive demands produced in employees by given tasks. This study laid the foundation for the development of personalized stress assistants. Physiological data of five groups of two participants were recorded, following a five-appointment study design. During the appointments, each pair underwent a cognitive load induction and subsequent stress reduction phase. Physiological signals were recorded with low-cost wearable sensors, subsequently analyzed for biomarkers, and compared for similarity between participants and groups. Results show that the sensors are capable of capturing descriptive data. Despite simultaneous task executions, it was found from the similarity analysis that the normalized Dynamic Time Warping distances between extracted features are greater for yoga sessions than during the cognitive load sessions. The classification of tasks was performed using the Machine Learning algorithms (i) Logistic Regression, (ii) Support Vector Machines, (iii) Nearest Neighbors, and (iv) Decision Trees trained on feature sets of either the Muse S, the Empatica E4, or both sensors together. Generalized as well as personalized models achieved classification accuracies over 85.00%. The recorded data is available upon request. The stimulus elicitation framework developed using PsychoPy and the software artifacts for data analysis were made publicly available, enabling the research community to evaluate their methods on this dataset and re-use analysis methods on their own or other datasets.**

Keywords—Mental Health; Mental Workload; Stress; Wearables; eHealth.

I. INTRODUCTION

To perform any natural task, humans utilize mental resources. In this context, a widely referenced concept is mental workload. According to [1], "*Mental workload may be viewed as the difference between the capacities of the information processing system that are required for task performance to satisfy performance expectations and the capacity available at any given time.*". It has been shown that the risk of coronary heart disease and hypertension, amongst other diseases, is increased if the mental workload is sustained at an elevated level over a long time, as mental workload alters the cardiovascular function, leading to a rising heart rate and blood pressure [2], [3].

To counteract such adverse consequences, these elevated levels of mental workload first need to be identified. Different avenues exist, such as performance-based, subjective, and physiological approaches. Performance-based measures mainly highlight situations where high levels of mental workload lead to mental overload. Subjective measures include self-assessments, but it has been shown that humans perform poorly in self-identifying decreased vigilance and cognitive overload [4]. Physiological measures are based on changes in the body incurred by mental workload, such as pupil dilation, heart rate, and changes in skin conductance. These measures can work on a continuous scale but usually require specialized equipment and trained staff [5].

A review on measuring mental workload covering Electrocardiogram (ECG), blood pressure, respiratory, ocular and dermal sensors alongside Electroencephalography (EEG), found that different measures can be used to discriminate task load, task type, and task difficulty while underlining the importance of multi-modal setups [6]. Furthermore, it was shown that one-channel in-ear EEG might suffice in optimal circumstances [7], while stress reduction can be predicted using ECG data from wearable sensors, amongst others [8]. As for mental workload, another interesting phenomenon was observed: by unconscious synchronization of brain activity across individuals, these individuals might utilize more mental resources than each individual alone would be able to [9]. This phenomenon was studied in various settings, such as communication [10] and learning processes between teachers and students, where the strength of the personal bond was found to be a modulator [11], [12].

To the best of the authors' knowledge, no related work focused on incorporating the analysis of group-wide processes of physiological signals in evaluating mental workload, stress, and stress-reduction interventions. Here, the reliability of wearable sensor systems on mental workload, stress, and activity type classification was investigated. Furthermore, a similarity analysis pipeline using the well-studied oddball paradigm [13] was validated, to quantify the effect of a yoga intervention in reducing mental workload and stress.

The remainder of the work is structured as follows: in Section II, related work is presented, while Section III details the methods employed in this work. Section IV gives the results of this work, for which future work is given in Section V. Finally, the conclusion is given in Section VI.

II. RELATED WORK

It was found that EEG measurements have adequate time resolution, conveying information online, and thus providing a promising tool for assessing cognitive workload, comparable in simplicity to measuring the physical workload with heart rate monitors or pedometers [4]. This finding was extended by another review of mental workload classification using wearable on-body devices, finding that EEG seems most promising and should be included in every multi-modal setup, as 'it is the only method that is directly related to mental workload' according to [14].

Given the negative effect of distress, interventions to reduce the stress levels of participants are plentiful. As such, studies have investigated the effects of exposure to music and nature sounds [15], mind-body connection courses designed to reduce anxiety [16], and multi-dimensional stress reduction interventions employing cognitive, somatic, dynamic, emotive and hands-on interventions [17], amongst many more. In addition, various literature reviews were conducted on this topic [18], [19]. Yoga and breathing exercises are widely known as a specific form of mindfulness, practiced in various forms for thousands of years. Numerous literature reviews synthesized some of the key findings for yoga on individuals concerning reductions in depression symptoms, stress and anxiety ratings, as well as the frequency of symptoms, such as headaches, particularly also in a short time frame after the onset of the intervention [20]–[22]. It was found that practices that include yoga asanas appear to be associated with improved regulation of the sympathetic nervous system and hypothalamic-pituitary-adrenal system [22].

In light of movement-based interventions, the contamination of physiological time series with movement artifacts needs to be considered. As for ocular artifacts (looking at instructive yoga videos in the present study), conflicting evidence was found. One work found that no substantial artifacts were present in mobile EEG readings, naturally except for frontal recording sites [23], and another work found that eye movements significantly distorted recordings from electrodes at frontal, temporal, and ear positions [24]. Both works agree, however, that artifacts are generally stronger in EEG bands of higher frequency. Automatic artifact tagging algorithms were proposed, to classify movement artifacts as emerging from loss of contact with the sensor, or from movement of the underlying tissue, as demonstrated on EEG data [25]. Recently, the current state of the art of movement artifact removal from EEG was summarized, finding that software and hardware solutions need to be utilized simultaneously, and recommending guidelines [26].

As for another modality, the Photoplethysmography (PPG), it was found that wavelet transforms as well as Kalman filters might be needed to remove unwanted artifacts from the data, mitigating the impact of artifacts [27]. With the rise of Machine Learning (ML) techniques, artifact detection has shifted to employ such measures as well, as demonstrated by unsupervised artifact identification in another modality recorded at the wrist: electrodermal activity [28].

While well-studied event elicitation tests exist, (such as the Oddball paradigm, which is widely used for the analysis of event-related potentials in schizophrenia patients [29]), and synchronization algorithms for wearable devices exist (e.g., [30]), measurements of synchronicity of event-related responses recorded with wearable sensors are rarely but effectively performed [13]. The utilization of similarity measures for physiological data has recently gained some attention, especially for clinical decision support systems [31], but has, to the best of the authors' knowledge, rarely been performed for simultaneously recorded physiological data from wearable devices.

III. METHODS

Many challenges come up when working to synchronously record data from multiple participants, potentially even more so with wearable sensors than with hard-wired clinical-grade devices. As experimenters are usually not trained clinicians, the sensor fit of wearable devices is often of poorer quality than any clinical counterpart, with participant movements worsening the signal quality as described. Furthermore, signal transmission is mostly performed via third-party apps without explicit support for synchronous data recordings, shielding the experimenters from working with proprietary communication channels, while hiding a lot of the complexity inherent in synchronous data channels and potentially performing data cleaning on the (asynchronously) recorded data. This can lead to reduced trust in the recorded data if it was wholly recorded synchronously, or if some sensor clock-drift occurred or samples were dropped and interpolated at another time. To enable the research community to perform synchronous recordings in a multi-sensor and multi-user setup, a technical feasibility study was conducted in this work, including the conceptualization, development, and validation of a novel technical recording framework.

A. Utilized Sensors

As wearable sensors, the widely utilized wearable devices Empatica E4 and Muse S were employed. The Empatica E4 is a wrist-worn device, which contains Photoplethysmography (PPG; sensor read-out used to measure changes in the blood volume pulse), Electrodermal Activity (EDA, skin conductance measure correlating to stress, mental workload, and emotional responses), and accelerometer sensors. The Muse S headband contains four Electroencephalogram (EEG; records changes of the brain's electrical activity) sensors placed according to the 10/20 international system. Two frontal electrodes (AF7 and AF8) rest on the forehead and two temporal electrodes (TP9 and TP10) rest behind the ears. A reference sensor is located at the center of the forehead (FpZ). Apart from EEG sensors, Muse S contains PPG, gyroscope, and accelerometer sensors. Both wearables are commercial off-the-shelf devices, which have been tested and certified for safety under various regulatory standards, such as FCC and CE. The data was collected from the devices via a newly developed recording platform, implemented in Python 3.9 and building on top of PyLSL [32], a Python interface to the Lab Streaming Layer (LSL)) as well as on top of the Empatica E4 streaming server for Windows. For each wearable sensor, a separate BLED112 Bluetooth Dongle had to be utilized. The source code of the recording framework has been made publicly available at [33].

B. Study Design

During the study, five groups of two participants underwent five recording sessions on individual days. Each recording session lasted approximately 90 minutes, split into welcoming the respective pair of participants and fitting the sensors, a stress induction phase of approximately 30 minutes, and a yoga intervention of approximately 30 minutes succeeding the phase of high mental workload. Before, in between, and after the activities, subjective questionnaire data was collected from the participants. However, the yoga practice has not been interrupted to collect questionnaire answers, and as such subjective mental state assessments were collected only before and after the yoga practice. As questionnaires, the Brunel Mood Scale Questionnaire (BRUMS-Q), Stanford Sleepiness Scale (SSS), Visual Analogue Scale to Evaluate Fatigue Severity (VAS-F), as well as five-point Likert scales in the dimensions of mental workload and stress were utilized.

The induction of mental workload and stress was realized using randomized assignments of the widely used mental workload tasks AX-Continuous Performance Task (AX-CPT) [34] and Time Load Dual Back Task (TloadDback) [35], implemented in Python and presented using the PsychoPy platform [36]. Figure 2 gives an overview of the cognitive load induction framework. For the intervention, a publicly available Yoga video [37] was reproduced on a 75-inch TV screen. Half of the recordings (12 sessions) took place in a controlled environment at the Hasso Plattner Institute Campus 3, House G2, in Potsdam, Germany, a well-illuminated room with floor-to-ceiling windows on two sides of the room offering a view to trees. The other half of the recordings took place in uncontrolled environments. Out of a variety of options, the homes of some of the participants were chosen as uncontrolled environments at the request of the participants. Repeating some yoga poses, a sequence of 29 asanas was performed and finished with Shavasana and a chant of Om. Figure 1 gives a schematic overview of the study design. The cognitive load induction is described in detail in Figure 2. After twenty trials, the performance was assessed. If less than 85% of correct responses were achieved, the system added 0.1 seconds to each Stimulus Time (ST) and Response Time (RT) and repeated the process of Individualization. However, if the user had achieved 85% performance or more, the framework moved on with the current ST and RT settings to the final task for the remaining duration of the cognitive load induction phase.

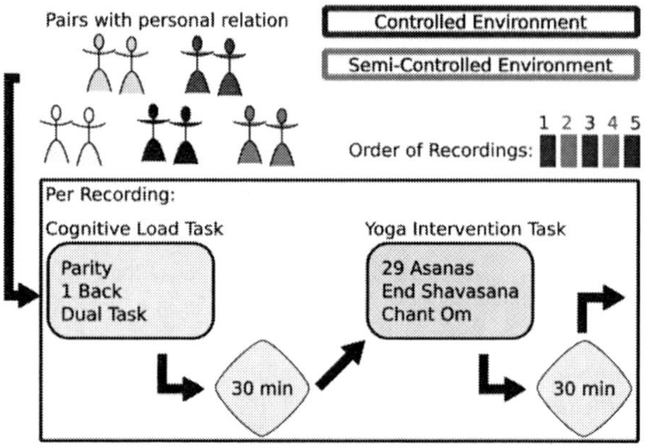

Figure 1: Overview of the study design. As for the Yoga intervention, 20 unique asanas were utilized by the video instructor (e.g., Child Pose, Cat and Cow, Downward-Facing Dog, etc.).

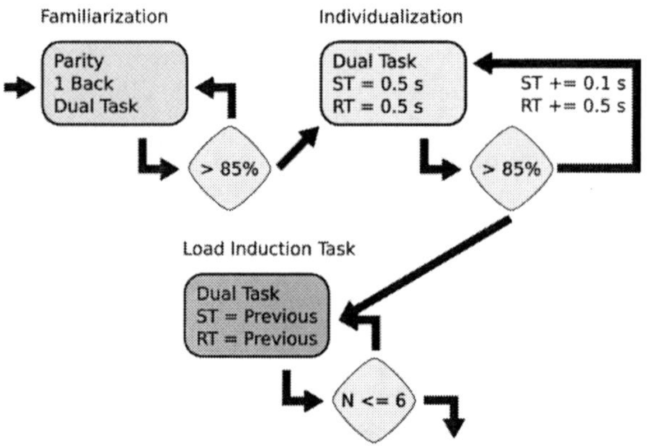

Figure 2: Overview of the cognitive load induction framework. Participants were first familiarized with the individual tasks. After a total of 60 trials, the participant's performance was assessed. If more than 85% of the cues were responded correctly, the user moved on to the individualization phase, which started directly with the lowest Stimulus Time (ST) and Response Time (RT).

Ethical approval has been obtained from the Institutional Review Board (IRB) of the University of Potsdam (application number 69/2023), and written informed consent was given by all participants before participating in the study. The study inclusion criteria required participants to be aged 18 to 33, sufficiently fluent in English (at least B2 level), have a normal or corrected-to-normal vision, know how to use a smartphone, and have to regularly perform work that was performance-evaluated (e.g., students or employees). Participants were required to regularly perform sports or yoga, to be experienced with moderate at-home workouts, stretching, and video-based yoga, and to be in a close relationship with the participant they registered with.

The study exclusion criteria excluded participants who needed to regularly take medication, such as mood stabilizers or psychotropic drugs and could not record data for approximately 90 minutes without interruptions except for bathroom visits.

As the study required the participants to perform specific yoga exercises, physically disabled or injured persons (recovered for less than six months) had to be excluded in case the prospective participant was unable to perform the majority of the required movements. Furthermore, participants who could have been in any dependent relationship with the experimenters, pregnant women, and participants with hypertension were excluded. Out of an overwhelming response to study recruitment efforts, a random total number of ten participants were recruited and recorded to evaluate the technical feasibility of the study setup.

C. Similarity

To confirm the synchrony of the recorded data, two sanity checks were integrated into the study protocol. Firstly, the experimenters vigorously shook the recording devices at the beginning and end of the recordings for approximately ten seconds. This ensured simultaneous peaks in the acceleration data of the wearables, and as a result enabled the comparison of peak onset and offset times, validating that no clock drift had occurred during the recordings and that by consequence the time series between the well-aligned peaks in the start and at the end of the recordings had to be well-aligned as well. Secondly, an Oddball paradigm was utilized to validate if it was possible to measure Steady-State Visual Evoked Potentials (SSVEPs) with the Muse S wearable EEG headband and to analyze the synchrony of these SSVEPs.

However, due to calibration issues with the TV screen, the majority of the Oddball paradigm sessions were not reproduced with the anticipated 60 Hz refresh rate of the screen and a matching signal rate, but with a refresh rate much lower, resulting in invalid Oddball recordings that had to be interrupted due to excessive durations and very slow signal changes. Due to Bluetooth data transmission and Bluetooth channel saturation, drops in sampling frequencies of the individual sensors occurred. Mostly, however, the Muse S sampled EEG data at 256 Hz, PPG data at 64 Hz, and Gyroscope and Acceleration data at 50 Hz. The Empatica E4 mostly sampled BVP data at 64 Hz, Acceleration data at 32 Hz, and GSR as well as Skin Temperature data at 4 Hz.

D. Data Processing

During data recording, the data was stored in .h5 format. After each recording session was stopped, the newly developed streaming platform StreamSense immediately triggered a data cleaning and data processing pipeline ProSense, creating signal quality reports and subsequently storing the recorded data in .pkl format. Figure 3 gives an abstract representation of the data preprocessing flow, triggered automatically after each recording session. The individual parameters, such as outlier rejection thresholds for the dynamic Interquartile Range (IQR) method, pass- and stop-band definitions, as well as the normalization method utilized (min-max), are documented in the source code documentation of ProSense. Alongside the sensor data, log files were created from Questionnaire answers, performance times, reaction times, and system logs. For each recording, the logs were cleaned, a processed subset was stored, and features were extracted and stored in individual .csv files corresponding to the respective modality.

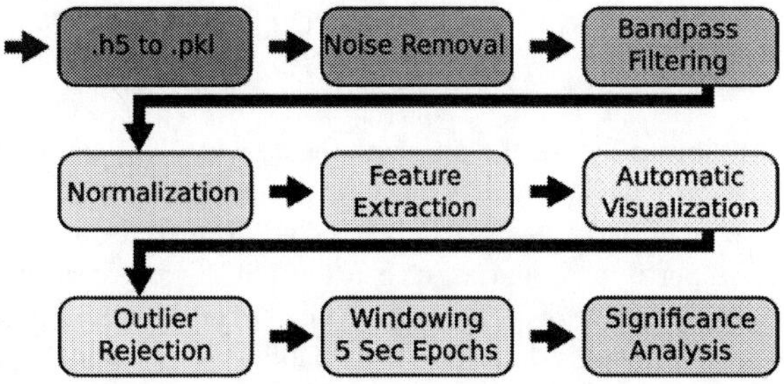

Figure 3: Overview of the data processing pipeline, triggered automatically by ProSense after each recording.

Across files, the same (anonymized) identifiers for participants as well as timestamps were utilized. Features that were extracted are Kurtosis, Skewness, Entropy, Min, Mean, and Max for Acceleration data, BVP data, and Gyroscope data, amongst others. For EEG data, the main features extracted were power spectral densities, band-powers, band ratios at the different electrodes, spectral entropy, and various statistical features. For the GSR data, the skin conductance level and the skin conductance response value were extracted, amongst others. For the PPG data, the heart rate, heart rate variability, and others were extracted. As a window length of features, an epoch duration of five seconds was utilized. The source code for the data storage and feature extraction was made publicly available at [38].

E. Machine Learning

As a final step, Machine Learning (ML) models were trained to distinguish between the activities performed by the study participants. As ML models, the widely used model-families Logistic Regression (LR), Decision Trees (DT), Nearest Neighbors (NN), and Support Vector Machines (SVM) were employed. Effectively, the ML models were trained as generalized binary activity classifiers. The hyperparameters for each ML model were derived using a nested 5-fold cross-validation scheme, training and evaluating the model performance for a given set of hyperparameters and testing the generalization capabilities on a held-out test set. Hyperparameters for the LR were penalty (*l1, l2, None*) and solver (*lbfgs, liblinear, sag, saga*), for the DT were criterion (*gini, entropy*), splitter (*best, random*, and max_depth (*5, 10, ..., 300, None*)), for the SVM (Linear Support Vector Classifier) were penalty (*l1, l2*), as well as the regularization parameter C (*0.01, 0.1, 1,*

10, 100, 1000), and for the NN were (leaf_size (*1, 2, ..., 50*), n_neighbors (*1, 2, ..., 30*), and p (*1, 2*)). The train-validate-test split was 60%-20%-20%, and as outer stratified 5-fold CV was employed, while the experimental HalvingGridSearchCV from scikit-learn was utilized for the inner CV [39].

Finally, the resulting performances were averaged and the best hyperparameters were noted down. Due to data imbalances, (*41%:59%* for *Cognitive-Load:Intervention*), the data was once randomly resampled before the experiments, resulting in balanced data sets.

IV. RESULTS

A. Machine Learning

The mean age of the ten participants was 27.6 years, with a standard deviation of 4.34 years. Due to the sickness of one pair of participants, their respective fifth recording could not be performed, and as such a total of 48 data recordings (24 sessions) were performed. After hyperparameter tuning utilizing nested 5-fold CV and HalvingGridSearchCV, the following hyperparameters were utilized across most of the model runs: for the LR (*penalty = None, solver = lbfgs*), for the DT (*criterion = entropy, splitter = best, max_depth = 145*), for the SVM (*penalty = l1, C = 1000*), and for the NN (*leaf_size = 25, n_neighbors = 21, p = 1*). The resulting model performances for distinguishing between cognitive load induction and yoga intervention are detailed in Table I. The mean across nested CVs (Generalized) or across nested CVs and across participants (Personalized) is reported. The feature sets utilized contained *Kurtosis, HRV, HR, SCL, SCR_Freq*, and *Temp* features (E4), *AF7_alpha_power, AF8_alpha_power, TP9_alpha_power, TP10_alpha_power, AF8_theta_delta, AF7_theta_delta, AF7_low_beta, AF8_low_beta, tfr_9Hz, tfr_18Hz, tfr_27Hz, entropy_AF8, entropy_AF7, entropy_TP9*, and *entropy_TP10* features (Muse), or all of these (All). As can be seen, both for Generalized and Personalized models, the NN (printed in boldface) performed best, while overall DT performed worst. The best performance was consistently achieved using the feature set All, followed by the Muse features, and finally by the E4 features.

An exemplary visualization of some features averaged across all participants is given in Figure 4. Feature values were averaged per participant across the min-max normalized values (and for the EEG features at the electrode positions AF7, AF8, TP9, and TP10), and averaged across recordings. As can be seen, the Alpha Power, which correlates positively with relaxation [40], is increased during the yoga intervention when compared to the cognitive load induction phase, validating its use as a biomarker for cognitive demands. While the heart rate does not seem to change significantly between conditions, the SCL is higher during the intervention than during the load induction at rest. The strong distortion in the physiological signals around the time of the transition from one phase to another, including a lot of uncontrolled movements, is reflected in the data.

Model	Feature-Set	Generalized	Personalized
NN	All	88.80%	**90.01%**
NN	Muse	84.28%	**86.59%**
NN	E4	72.64%	**79.68%**
LR	All	80.12%	**82.13%**
LR	Muse	68.94%	**80.77%**
LR	E4	73.36%	**73.81%**
SVM	All	79.73%	**81.33%**
SVM	Muse	68.40%	**80.45%**
SVM	E4	73.32%	**73.98%**
DT	All	78.06%	**78.60%**
DT	Muse	71.62%	**79.51%**
DT	E4	68.42%	**75.75%**

Table I: Classification accuracy for binary classification models on well-balanced (*50%:50%* datasets for *cognitive load:yoga*), derived after nested 5-fold Cross-Validations (CV). Generalized models were built using data from all participants while personalized models utilized data from only one participant (no outliers removed). Overall model performances are color-coded from best (blue) to worst (red). The best performance per row is printed in boldface.

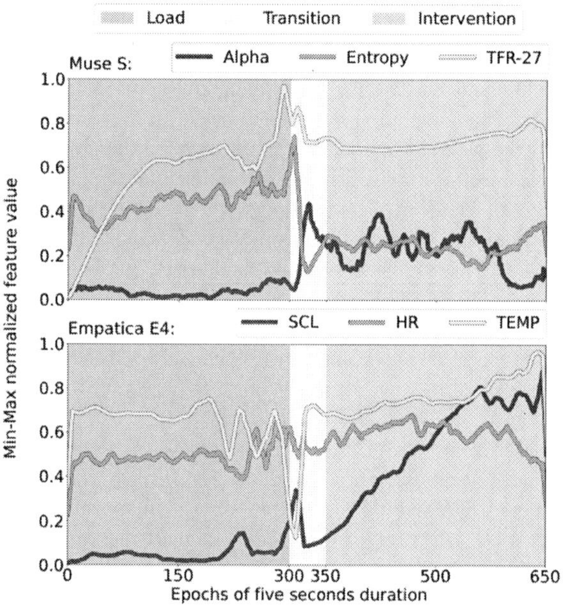

Figure 4: Generalized mean feature values for mean power in the *alpha frequency band (8 - 12 Hz: Alpha), spectral entropy (Entropy)*, and the *spectral power at 27 Hz (TFR-27)* derived for the Muse S (top), and the *skin conductance level (SCL), heart rate (HR)*, and *skin temperature (TEMP)* derived for the Empatica E4 (bottom). The signal was smoothed over twelve consecutive epochs of five seconds, i.e., over one minute.

B. Similarity Analysis

To confirm the validity of synchrony of the recorded EEG data, initially a comparison of the SSVEPs after Oddball paradigm had been planned. However, due to the issues outlined in the Subsection *Similarity*, only two recordings out of the total of 48 data recordings could be considered for the analysis of Event-Related Potentials (ERPs). Figure 5 visualizes these results, which are not generalizable as the analysis was performed only on a few data points. Due to technical difficulties with the oddball presentation paradigm outlined in the Subsection *Similarity*, the data shown is averaged over one session of two participants, respectively. The well-studied ERP components N200 and P300 are well-visible for the oddball paradigm. While the absence of the P300 in the *control* task is expected [13], it is unexpected that no N200 ERP is visible. As a result of the technical difficulties, the absence of the N200 in the *control* task, and the low number of samples, the reliability of ERP analysis on this data is limited. However, in line with related work [13], these results underline the possibility of researching SSVEPs with the utilized low-cost sensors.

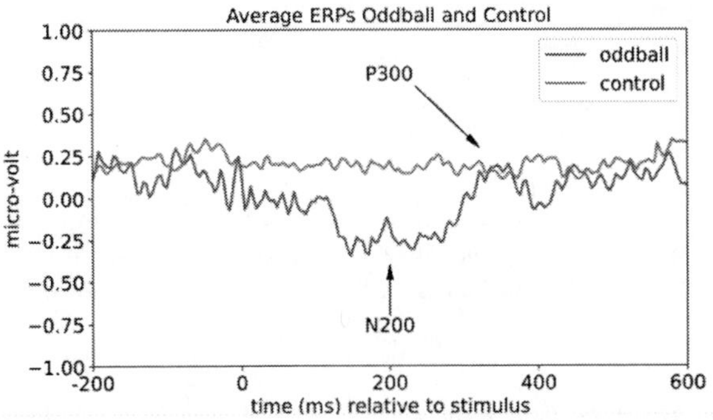

Figure 5: Event-Related Potential (ERP) analysis after oddball paradigm during *oddball (blue)* and *control (green)* tasks, respectively.

Another approach to analyzing the similarity of the recorded physiological signals is utilizing a distance-based measure for the features extracted from the physiological data [41]. In this work, a Python implementation for Dynamic Time Warping (DTW) was utilized [42]. Compared to some other similarity measures, DTW allows for non-linear matching by stretching or shrinking the compared signals [43], and has also been explored in ML [44]. One challenge for this analysis is that the participants were instructed via video-based yoga to perform the same movements. Consequently, during the recordings, participants did not necessarily perform the same exercise the same way at the same time after an instructed change of pose. A non-linear distance-based measure, such as DTW, is well-suited for this analysis [42]. Here, the normalized DTW distance across the Empatica E4 feature sets *Kurtosis, HRV, HR, SCL, SCR_Freq*, and *Temp* features, and across the Muse S feature sets *AF7_alpha_power, AF8_alpha_power, TP9_alpha_power, TP10_alpha_power, AF8_theta_delta, AF7_theta_delta, AF7_low_beta, AF8_low_beta, tfr_9Hz, tfr_18Hz, tfr_27Hz, entropy_AF8, entropy_AF7, entropy_TP9*, and *entropy_TP10*, was computed between each epoch of each recording. Special interest was placed on enabling the comparison between the pairs of participants. The results are visualized in Figure 6. The color-coded boxes represent the distance within a group of participants, across all sessions. The white box represents the distance of all participants not within the same group, across all sessions. Boxes start at the mean distances during cognitive load and yoga sessions, and their width and height are given by the respective standard deviations. As can be seen, the distances within the groups are smaller than between participants from different groups, but with a high standard deviation. Across participants and groups, the Standard Deviation (STD) of the mean normalized distance across all features and epochs is smaller than the STD over all Muse S features. Generally, the distances within the groups are smaller than the distances between the individuals of the respective group and other recordings.

C. Feature Importance

The importance of individual features was investigated using a correlation analysis performed after artifact removal. To remove the artifacts, a dynamic Interquartile Range (IQR) method built on the STD in each feature was utilized. Details can be found in the source code at [38]. Especially the statistical features extracted from the EEG data (correlation over 0.58 at p-values under 0.001), the heart rate variability (correlation of 0.51 at p-value under 0.001), and the skin conductance level (correlation of 0.45 at p-value under 0.001) were highly correlated with the phase.

D. Limitations

As the technical framework was constantly developed once a bug or a sub-optimal solution was noticed, some recordings produced slightly different artifacts than others. As a result thereof and of issues encountered in the uncontrolled environments, such as a vast amount of Bluetooth devices present in the immediate neighborhood, three recordings show a significant amount of artifacts, and one out of these recordings stored data for all modalities only at a maximum sampling rate of 10 Hz. Generally, due to the nature of the bodily exercise of yoga, the second half of the recordings is partially distorted due to strong movement artifacts when participants changed their yoga poses (only during said change). Furthermore, data labelling during yoga was impractical, as it would have interfered with the participants performing the stress-reducing intervention. Consequently, the temporal resolution of self-assessed labels is significantly higher for the cognitive load task than for the relaxation intervention task. Finally, the recordings were performed in winter, and some participants reported feeling a bit sick. Therefore, some participants asked for the windows to be closed, while other participants appreciated open windows, potentially influencing the comparability of temperature and GSR readings across recordings.

V. FUTURE WORK

Due to the richness of the dataset collected, some aspects remain to be analyzed further. The synchronicity of physiological responses during cognitive load induction, but especially

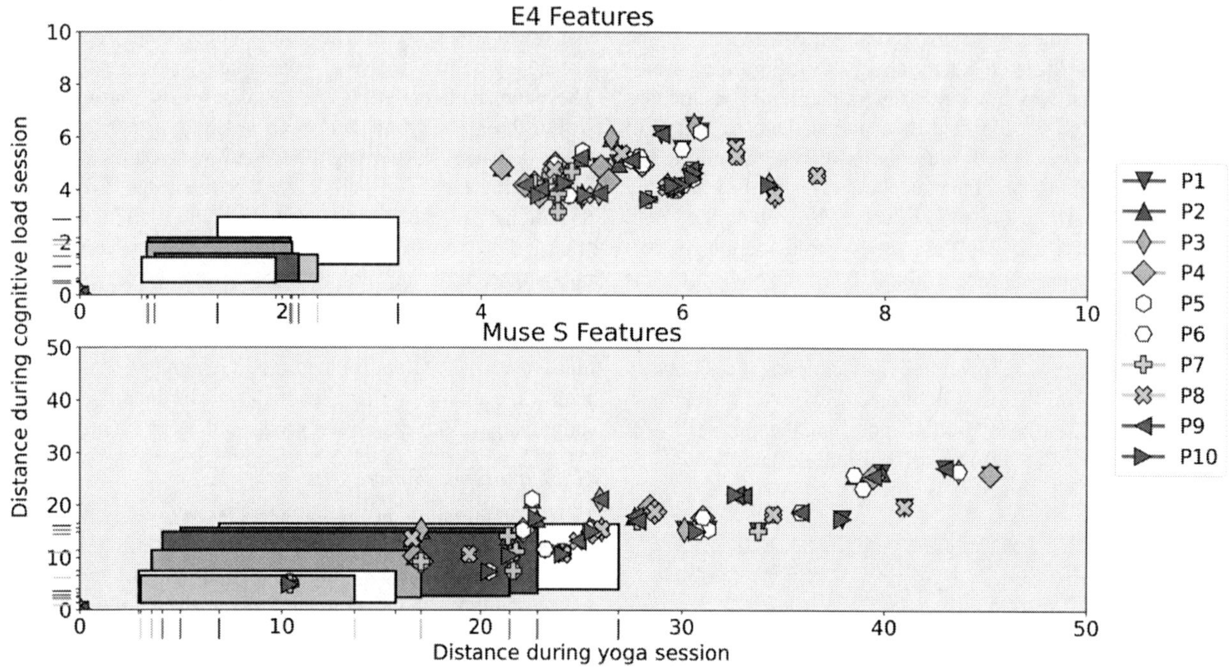

Figure 6: Mean normalized Dynamic Time Warping distances across features from the Empatica E4 (top) and Muse S (bottom), respectively [42]. The pairings of distances are given for each session for the participants not within the same group (i.e., session one of participant P1 was compared to all the 1st sessions of all other participants but the group-partner of P1) and labeled with the markers in the legend.

during the stress reduction mechanisms, should be investigated further. By performing subsequent data collection using the same protocol on individuals rather than on small groups, the stress reduction as determined by the biomarkers could be analyzed and compared, potentially leading to tangible recommendations for organizations' policies. Personalizing the analysis even further, it would be possible to conduct the same analyses and ML regressions using the participant-given labels. If the binary classifiers were trained on the actual user-perceived labels and not on predefined task labels (data available), the classification results are expected to be different. Lastly, ML and Deep Learning models existing in related work could be further personalized on this dataset, and the resulting models could be made publicly available while investigating the effective usefulness of ML and DL compared with traditional statistics.

VI. CONCLUSION

This study's findings on biomarkers of cognitive demands and their ease of use for ML classifiers have significant implications for Personalized eHealth, particularly regarding the development of personalized stress management solutions. Physiological data of five groups of two participants were recorded, following a five-appointment study design. During the appointments, each pair underwent a cognitive load induction and subsequent stress reduction phase. Results show that the sensors are capable of capturing descriptive data. Despite simultaneous task executions, it was found from the similarity analysis that the normalized Dynamic Time Warping distances between extracted features are greater for yoga sessions than during the cognitive load sessions. The derived load classifiers can be integrated into eHealth platforms and offer monitoring or tailored advice on interventions based on the individuals' stress patterns. As such, real-time stress detection would enable immediate suggestions of coping mechanisms like guided breathing exercises or mindfulness meditation prompts. Moreover, the rich dataset of this study, available upon request, offers immense potential for advancing the understanding of stress physiology in real-world applications, which can be leveraged to refine eHealth technologies further, ensuring they meet the unique needs of each individual. This personalized approach not only enhances user engagement but also promises improved health outcomes by addressing stress in a timely and relevant manner and could therefore help shift organizations towards an employee-focused workplace.

ACKNOWLEDGMENT

This research was (partially) funded by the HPI Research School on Data Science and Engineering.

REFERENCES

[1] D. Gopher and E. Donchin, "Workload: An examination of the concept," in *Handbook of perception and human performance, Vol. 2: Cognitive processes and performance*. John Wiley & Sons, pp. 1–49, retrieved: 4, 2024. [Online]. Available: https://psycnet.apa.org/record/1986-98619-019

[2] S. G. Hart and L. E. Staveland, "Development of NASA-TLX (task load index): Results of empirical and theoretical research," in *Advances in Psychology*, ser. Human Mental Workload. North-Holland, vol. 52, pp. 139–183, retrieved: 4, 2024. [Online]. Available: https://www.sciencedirect.com/science/article/pii/S0166411508623869

[3] S. Delliaux, A. Delaforge, J.-C. Deharo, and G. Chaumet, "Mental workload alters heart rate variability, lowering non-linear dynamics," vol. 10, retrieved: 4, 2024. [Online]. Available: https://www.frontiersin.org/article/10.3389/fphys.2019.00565

[4] A. Holm, K. Lukander, J. Korpela, M. Sallinen, and K. M. I. Müller, "Estimating brain load from the EEG," vol. 9, pp. 639–651, retrieved: 4, 2024. [Online]. Available: http://www.hindawi.com/journals/tswj/2009/973791/abs/

[5] L. Longo, F. Rusconi, and L. Noce, "The importance of human mental workload in web design," in *Proceedings of the 8th International Conference on Web Information Systems and Technologies*. SciTePress - Science and and Technology Publications, pp. 403–409, retrieved: 4, 2024. [Online]. Available: http://www.scitepress.org/DigitalLibrary/Link.aspx?doi=10.5220/0003960204030409

[6] R. L. Charles and J. Nixon, "Measuring mental workload using physiological measures: A systematic review," vol. 74, pp. 221–232, retrieved: 4, 2024. [Online]. Available: https://www.sciencedirect.com/science/article/pii/S0003687018303430

[7] J. M. Morales, J. F. Ruiz-Rabelo, C. Diaz-Piedra, and L. L. Di Stasi, "Detecting mental workload in surgical teams using a wearable single-channel electroencephalographic device," vol. 76, no. 4, pp. 1107–1115, retrieved: 4, 2024. [Online]. Available: https://www.sciencedirect.com/science/article/pii/S1931720418306597

[8] A. Tonacci *et al.*, "Can machine learning predict stress reduction based on wearable sensors' data following relaxation at workplace? a pilot study," vol. 8, no. 4, p. 448, number: 4 Publisher: Multidisciplinary Digital Publishing Institute. [Online]. Available: https://www.mdpi.com/2227-9717/8/4/448

[9] Y. Pan, X. Cheng, and Y. Hu, "Three heads are better than one: cooperative learning brains wire together when a consensus is reached," vol. 33, no. 4, pp. 1155–1169, retrieved: 4, 2024. [Online]. Available: https://doi.org/10.1093/cercor/bhac127

[10] A. Kuhlen, C. Allefeld, and J.-D. Haynes, "Content-specific coordination of listeners' to speakers' EEG during communication," vol. 6, retrieved: 4, 2024. [Online]. Available: https://www.frontiersin.org/articles/10.3389/fnhum.2012.00266

[11] S. Dikker *et al.*, "Brain-to-brain synchrony tracks real-world dynamic group interactions in the classroom," vol. 27, no. 9, pp. 1375–1380, retrieved: 4, 2024. [Online]. Available: https://linkinghub.elsevier.com/retrieve/pii/S0960982217304116

[12] D. Bevilacqua *et al.*, "Brain-to-brain synchrony and learning outcomes vary by student–teacher dynamics: Evidence from a real-world classroom electroencephalography study," vol. 31, no. 3, pp. 401–411, retrieved: 4, 2024. [Online]. Available: https://doi.org/10.1162/jocn_a_01274

[13] O. E. Krigolson *et al.*, "Using muse: Rapid mobile assessment of brain performance," vol. 15, retrieved: 4, 2024. [Online]. Available: https://www.frontiersin.org/journals/neuroscience/articles/10.3389/fnins.2021.634147

[14] C. Marchand, J. B. De Graaf, and N. Jarrassé, "Measuring mental workload in assistive wearable devices: a review," vol. 18, no. 1, p. 160, retrieved: 4, 2024. [Online]. Available: https://doi.org/10.1186/s12984-021-00953-w

[15] E. Largo-Wight, B. K. O'Hara, and W. W. Chen, "The efficacy of a brief nature sound intervention on muscle tension, pulse rate, and self-reported stress: Nature contact micro-break in an office or waiting room," vol. 10, no. 1, pp. 45–51, retrieved: 4, 2024. [Online]. Available: https://doi.org/10.1177/1937586715619741

[16] C. Finkelstein, A. Brownstein, C. Scott, and Y.-L. Lan, "Anxiety and stress reduction in medical education: an intervention," vol. 41, no. 3, pp. 258–264, retrieved: 4, 2024. [Online]. Available: https://onlinelibrary.wiley.com/doi/abs/10.1111/j.1365-2929.2007.02685.x

[17] S. Sallon, D. Katz-Eisner, H. Yaffe, and T. Bdolah-Abram, "Caring for the caregivers: Results of an extended, five-component stress-reduction intervention for hospital staff," vol. 43, no. 1, pp. 47–60, retrieved: 4, 2024. [Online]. Available: https://doi.org/10.1080/08964289.2015.1053426

[18] M. Sharma and S. E. Rush, "Mindfulness-based stress reduction as a stress management intervention for healthy individuals: A systematic review," vol. 19, no. 4, pp. 271–286, retrieved: 4, 2024. [Online]. Available: https://doi.org/10.1177/2156587214543143

[19] S. A. Smith, "Mindfulness-based stress reduction: An intervention to enhance the effectiveness of nurses' coping with work-related stress," vol. 25, no. 2, pp. 119–130, retrieved: 4, 2024. [Online]. Available: https://onlinelibrary.wiley.com/doi/abs/10.1111/2047-3095.12025

[20] E. Della Valle *et al.*, "Effectiveness of workplace yoga interventions to reduce perceived stress in employees: A systematic review and meta-analysis," vol. 5, no. 2, p. E33, retrieved: 4, 2024. [Online]. Available: https://doi.org/10.3390/jfmk5020033

[21] D. Anheyer, P. Klose, R. Lauche, F. J. Saha, and H. Cramer, "Yoga for treating headaches: a systematic review and meta-analysis," vol. 35, no. 3, pp. 846–854, retrieved: 4, 2024. [Online]. Available: https://doi.org/10.1007/s11606-019-05413-9

[22] M. C. Pascoe, D. R. Thompson, and C. F. Ski, "Yoga, mindfulness-based stress reduction and stress-related physiological measures: A meta-analysis," vol. 86, pp. 152–168, retrieved: 4, 2024. [Online]. Available: https://doi.org/10.1016/j.psyneuen.2017.08.008

[23] D. Hagemann and E. Naumann, "The effects of ocular artifacts on (lateralized) broadband power in the EEG," vol. 112, no. 2, pp. 215–231, retrieved: 4, 2024. [Online]. Available: https://doi.org/10.1016/s1388-2457(00)00541-1

[24] S. L. Kappel, D. Looney, D. P. Mandic, and P. Kidmose, "Physiological artifacts in scalp EEG and ear-EEG," vol. 16, no. 1, p. 103, retrieved: 4, 2024. [Online]. Available: https://doi.org/10.1186/s12938-017-0391-2

[25] K. T. Sweeney, D. J. Leamy, T. E. Ward, and S. McLoone, "Intelligent artifact classification for ambulatory physiological signals," vol. 2010, pp. 6349–6352, retrieved: 4, 2024. [Online]. Available: https://doi.org/10.1109/IEMBS.2010.5627285

[26] D. Gorjan, K. Gramann, K. De Pauw, and U. Marusic, "Removal of movement-induced EEG artifacts: current state of the art and guidelines," vol. 19, no. 1, retrieved: 4, 2024. [Online]. Available: https://doi.org/10.1088/1741-2552/ac542c

[27] W.-J. Lin and H.-P. Ma, "A physiological information extraction method based on wearable PPG sensors with motion artifact removal," in *2016 IEEE International Conference on Communications (ICC)*, pp. 1–6, ISSN: 1938-1883. [Online]. Available: https://ieeexplore.ieee.org/document/7511485

[28] Y. Zhang, M. Haghdan, and K. S. Xu, "Unsupervised motion artifact detection in wrist-measured electrodermal activity data," in *Proceedings of the 2017 ACM International Symposium on Wearable Computers*, ser. ISWC '17. Association for Computing Machinery, pp. 54–57, retrieved: 4, 2024. [Online]. Available: https://doi.org/10.1145/3123021.3123054

[29] B. Alejandro *et al.*, "A comparative study of event-related coupling patterns during an auditory oddball task in schizophrenia," vol. 12, no. 1, p. 016007, retrieved: 4, 2024. [Online]. Available: https://iopscience.iop.org/article/10.1088/1741-2560/12/1/016007

[30] Z. Yan, R. Tan, Y. Li, and J. Huang, "Wearables clock synchronization using skin electric potentials," vol. 18, no. 12, pp. 2984–2998, retrieved: 4, 2024. [Online]. Available: https://ieeexplore.ieee.org/document/8565988

[31] C. Comito, D. Falcone, and A. Forestiero, "Diagnosis detection support based on time series similarity of patients physiological parameters," in *2021 IEEE 33rd International Conference on Tools with Artificial Intelligence (ICTAI)*, pp. 1327–1331, ISSN: 2375-0197. [Online]. Available: https://ieeexplore.ieee.org/document/9643300

[32] C. Kothe, B. Venthur, C. Boulay, D. Medine, C. Brunner, and M. Grivich, "Python interface to the lab streaming layer (lsl)," https://github.com/chkothe/PyLSL, accessed May 22, 2024.

[33] S. S. Gadamsetti, "Streamsense," https://github.com/siddhant61/StreamSense, accessed May 22, 2024.

[34] D. Umbricht *et al.*, "Effects of the 5-HT2a agonist psilocybin on mismatch negativity generation and AX-continuous performance task: Implications for the neuropharmacology of cognitive deficits in schizophrenia," vol. 28, no. 1, pp. 170–181, retrieved: 4, 2024. [Online]. Available: https://www.nature.com/articles/1300005

[35] K. O'Keeffe, S. Hodder, and A. Lloyd, "A comparison of methods used for inducing mental fatigue in performance research: individualised, dual-task and short duration cognitive tests are most effective," vol. 63, no. 1, pp. 1–12, retrieved: 4, 2024. [Online]. Available: https://www.tandfonline.com/doi/full/10.1080/00140139.2019.1687940

[36] J. Peirce *et al.*, "PsychoPy2: Experiments in behavior made easy," *Behavior Research Methods*, vol. 51, no. 1, pp. 195–203, Feb.

2019, retrieved: 4, 2024. [Online]. Available: https://doi.org/10.3758/s13428-018-01193-y

[37] Kassandra, "30 min beginner yoga," https://www.youtube.com/watch?v=6hZIzMpHl-c, accessed May 22, 2024.

[38] S. S. Gadamsetti, "Prosense," https://github.com/siddhant61/ProSense, accessed May 22, 2024.

[39] F. Pedregosa *et al.*, "Scikit-learn: Machine learning in Python," *Journal of Machine Learning Research*, vol. 12, pp. 2825–2830, 2011, retrieved: 4, 2024. [Online]. Available: https://scikit-learn.org/stable/about.html

[40] K. Mathewson *et al.*, "Regional electroencephalogram (EEG) alpha power and asymmetry in older adults: a study of short-term test–retest reliability," vol. 7, retrieved: 4, 2024. [Online]. Available: https://www.frontiersin.org/articles/10.3389/fnagi.2015.00177

[41] E. Keogh and C. A. Ratanamahatana, "Exact indexing of dynamic time warping," vol. 7, no. 3, pp. 358–386, retrieved: 4, 2024. [Online]. Available: http://link.springer.com/10.1007/s10115-004-0154-9

[42] T. Giorgino, "Computing and visualizing dynamic time warping alignments in r: The dtw package," vol. 31, pp. 1–24, retrieved: 4, 2024. [Online]. Available: https://doi.org/10.18637/jss.v031.i07

[43] S. Salvador and P. Chan, "Toward accurate dynamic time warping in linear time and space," vol. 11, no. 5, pp. 561–580, retrieved: 4, 2024. [Online]. Available: https://www.medra.org/servlet/aliasResolver?alias=iospress&doi=10.3233/IDA-2007-11508

[44] S. Gudmundsson, T. P. Runarsson, and S. Sigurdsson, "Support vector machines and dynamic time warping for time series," in *2008 IEEE International Joint Conference on Neural Networks (IEEE World Congress on Computational Intelligence)*. IEEE, pp. 2772–2776, retrieved: 4, 2024. [Online]. Available: http://ieeexplore.ieee.org/document/4634188/

Using Computer Vision-based Markerless Pose Estimation for Measuring Shoulder Range of Motion

Thomas Hellstén and Jonny Karlsson
Arcada University of Applied Sciences
School of Engineering, Culture and Wellbeing, Helsinki, Finland
emails: thomas.hellsten@arcada.fi, jonny.karlsson@arcada.fi

Abstract— **The use of health technology applications has increased during recent years among health care professionals. A novel and innovative approach for implementing health technologies in daily practice is through Computer Vision (CV) based markerless pose estimation. This approach is useful especially in rehabilitation applications for providing automatic guidance for clients performing rehabilitation exercises. The aim of this paper is to present the technical realization and early stage testing results of an open source prototype application for shoulder Range of Motion (ROM) analysis for rehabilitation purposes. The testing process included early stage accuracy tests of the prototype, in comparison to using a universal goniometer, for measuring all four active motion movements of the shoulder (flexion, extension, abduction, adduction). The results indicated that CV-based markerless pose estimation has the potential to accurately analyze shoulder joint ROM. In conclusion, the markerless CV application used in this study was found to have potential to be used in clinical practice by healthcare professionals. However, more comprehensive testing is still needed before it can be put into practice.**

Keywords- computer vision; range of motion; telerehabilitation; YOLO.

I. INTRODUCTION

The use of health technologies has been more frequent in recent years due to the COVID-19 pandemic, and enforced health care organizations to integrate Telerehabilitation (TR) into daily routines in clinical work [1]. Also, ensuring convenient and equitable access to health care services poses a notable challenge, given factors such as the aging population, rising incidence of chronic diseases, and the centralization of health, rehabilitation, and social services in urban areas [2]. TR is defined as a health care service that is delivered to clients through Information and Communication Technology (ICT) [3]. In physiotherapy, TR enables clients and health care professionals (physiotherapists) undergoing rehabilitation to connect from various locations, and stay in contact in real-time or asynchronous communication through ICT. However, it can also mean health technology applications that gives automatic information and support for the client [4].

There is evidence suggesting that TR could be as effective as and comparable to traditional physiotherapy in various diseases, such as rehabilitation following Parkinson's disease [5], heart diseases [6], non-specific chronic low back pain [7], stroke [8] and hip arthroplasty [9]. A benefit compared to traditional physiotherapy is that clients undergoing rehabilitation do not have to travel for TR, which increases the accessibility, whether influenced by the geographical location of clients or constraints within health care services [10]. There is also some evidence that TR is a more cost-effective method than traditional physiotherapy at a clinic [7].

A novel and innovative approach for implementing TR in daily practice in rehabilitation is through Computer Vision (CV) based markerless pose estimation. Pose estimation systems provide keypoint detection which can be utilized for detecting and locating joints of the human body. Based on estimated joint coordinates for each frame, different types of joint Range of Motion (ROM) assessment can be automatically performed.

Tracking and analysis of human movements using CV have been an important research topic for years [11]. CV typically employs marker-based methods, requiring placement of markers as reflective material on crucial body points (key points) like finger, elbow, shoulder and hip joints [12]. This limitation reduces the regular use of CV motion analysis systems impractical, as it requires extensive technical preparations before the use.

The benefit of using markerless pose estimation is that the only technical equipment needed is a regular computing device, such as laptop or smartphone, equipped with a camera. Several markerless pose estimation systems have been proposed and evaluated for rehabilitation purposes [13]. Most systems provide 2D joint detection using a single external web camera setup. Some solutions also provide 3D joint detection, by utilizing multiple camera views, providing the possibility to perform more advance ROM assessments.

However, markerless pose estimation systems bring some challenges that must be addressed before they can be integrated into rehabilitation practices. The main challenge is the necessity to establish sufficient accuracy of the measured data [14].

The aim of this paper is to introduce a novel markerless CV-based prototype application for measuring shoulder joint ROM for rehabilitation purposes and to preliminary validate its accuracy through early stage testing. The prototype requires only a single web camera and an off-the shelf laptop or desktop computer. The early stage testing procedure has been committed to an interdisciplinary research team, comprising experts from both physiotherapy and information technology. The structure of the rest of the paper is as follows. Section II introduces the new CV-based markerless prototype application and its technical features. Section III describes the early stage testing process and preliminary results. In Section IV, strengths and limitations of the prototype are discussed. Finally, some conclusions and future research directions are presented in Section V.

II. Prototype Application for Computer Vision-Based Markerless Shoulder Joint Range of Motion Measurement

A new prototype application for measuring shoulder joint ROM using a CV-based markerless technology was developed with both measurement accuracy and user accessibility taken into priority. To be able to achieve reliable ROM analysis, the shoulder, elbow, and hip joints must be located accurately. For detecting human joints, without using physical markers, a deep learning based key point detection methodology is needed which is typically computationally resource intensive.

You Only Look Once version 8 (YOLOv8) [15], however, is a popular object detection technique due to both its accuracy and speed which makes it advantageous for real time detection on a wide range of computation devices. YOLOv8 includes a pose estimation model, pretrained on the COCO dataset [16], that provides human body keypoint detection.

The keypoint detection feature of YOLOv8 was utilized in the prototype application for detecting and localizing the hip, shoulder and elbow joints for each frame. Hence, shoulder ROM analysis is performed by measuring and analyzing the shoulder angle α, Figure 1.

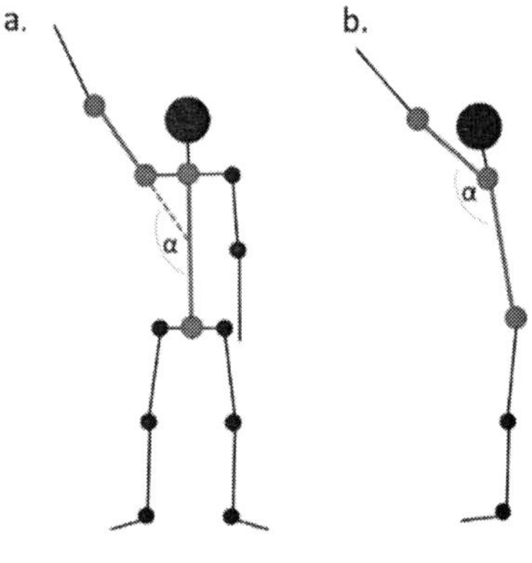

Figure 1. The shoulder angle α measured by the prototype a) from the front and b) from the side.

When assessing shoulder abduction and adduction, the client is standing with the front towards the camera (see Figure 2) and α is defined as the angle between the shoulder-elbow line and the midline of the body. The start and end point of the midline are estimated by calculating the center point between the shoulder and the hip joint pairs.

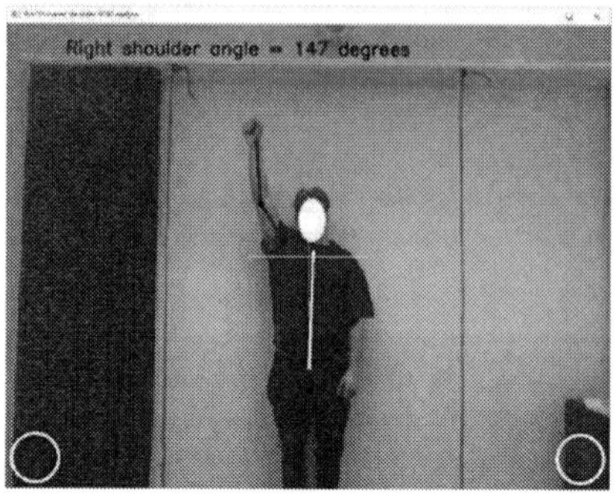

Figure 2. A screenshot of the prototype application when performing shoulder abduction and adduction assessment.

Correspondingly, when assessing shoulder flexion and extension the client stands with either side facing towards the camera, see Figure 3.

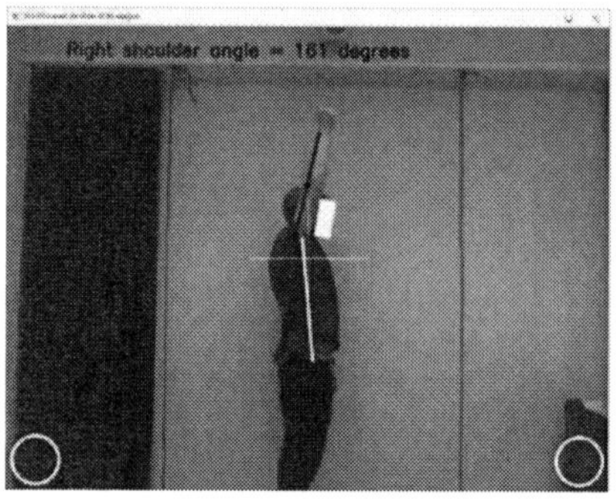

Figure 3. A screenshot of the prototype application when performing shoulder flexion assessment.

The shoulder angle α is in case of shoulder flexion/extension defined as the angle between the shoulder-elbow line and the shoulder-hip line.

III. Early Stage Testing

Early stage testing was conducted to compare the prototype application with Universal Goniometer (UG) when measuring shoulder joint ROM (flexion, extension, abduction and adduction) in standing position. UG was chosen as a reference for our CV prototype application as it is the most frequently used method by healthcare professionals for this purpose [17]. We followed the instructions by [18] when using bony landmarks for the UG measurement. UG measurements were performed by an experienced physiotherapist. The shoulder joint measurements were conducted in a standardized order: active flexion, active extension, active abduction and active adduction. Environmental factors were standardized as follows: White background and bright lightning were used in the test room. The distance between the test participant and the integrated laptop web camera was 2.85 m and the camera was positioned 1.35 m above the floor level. The test room and setup are shown in Figure 4.

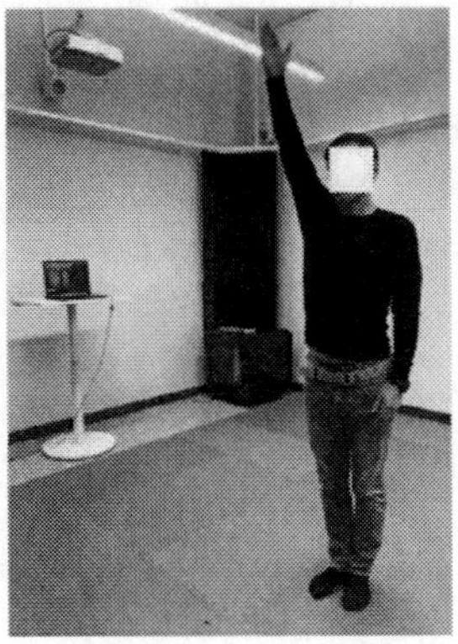

Figure 4. Picture of test setup when a participant performs active flexion in shoulder joint.

Three healthy voluntary persons over the age of 18 participated in the early stage test. All measurements were documented in a blind way, and each joint angle, measured by the prototype application, was stored in a log file with a timestamp. This information remained confidential until the corresponding angles were measured manually with UG. The angles were measured in whole degrees in the following order: 1) active shoulder flexion, 2) active shoulder extension, 3) active shoulder abduction and 4) active shoulder adduction.

All measured angles and measurement differences between the prototype application and UG are documented in Table I.

TABLE I. MEASURED ANGLES OF ACTIVE SHOULDER RANGE OF MOTION AND MEASUREMENT DIFFERENCES BETWEEN THE COMPUTER VISION (CV) PROTOTYPE APPLICATION AND UNIVERSAL GONIOMETER (UG)

Movement direction and difference	**Person 1**	**Person 2**	**Person 3**
Active shoulder flexion[a], CV	169	171	172
Active shoulder flexion[a], UG	170	170	175
Difference[a]	-1	1	-3
Active shoulder extension[a], CV	57	55	65
Active shoulder extension[a], UG	53	55	59
Difference[a]	4	0	6
Active shoulder abduction[a], CV	163	172	164
Active shoulder abduction[a], UG	170	165	170
Difference[a]	-7	7	-6
Active shoulder adduction[a], CV	17	14	17
Active shoulder adduction[a], UG	20	20	17
Difference[a]	-3	-6	0

[a]Values in degrees

IV. DISCUSSION

This paper has introduced a novel CV-based markerless prototype application for shoulder ROM analysis and presented preliminary accuracy evaluation results based on early stage testing. The prototype application performed the measurements with a moderate level of accuracy in all four active motion movements (flexion, extension, abduction, adduction).

Even though the measurements performed by the prototype application were not exactly equivalent with the manual UG measurements in this early stage test, the results can be considered promising. This, due to the fact that UG, that is the mostly used measurement tool by healthcare professional in daily practice, has showed a measure error of 6°. The use of radiograph images is the most accurate method for measuring the joint angle [19], however, due to ethical reason and high costs, this method was not used.

We recognize that the lightning in the room where the measurements were conducted influenced the results and how well the joints were detected by our prototype application (accuracy tends to decrease in darker environments).

V. CONCLUSION AND FUTURE WORK

CV is a promising technique for TR as it can be used for human motion analysis and show results to clients in real time without the need for complicated equipment. However, the accuracy of the joint detection capabilities of the motion analysis system is crucial in order to be able to provide a reliable motion analysis.

In this paper we have introduced a novel CV-based prototype application for measuring shoulder joint ROM. Early stage testing results indicated that our novel prototype application, using a CV based markerless pose estimation model, has potential for providing reliable shoulder ROM analysis. However, before a reliable accuracy level can be confirmed, and before CV based markerless pose estimation ROM analysis can be integrated in clinical practice, more comprehensive tests must be conducted. Therefore, future work should focus on testing the accuracy of the prototype application by performing measurements on a large group of persons, both healthy test persons but also on persons with different types of upper limb symptoms. Furthermore, tests must be conducted in environments with variable backgrounds and lightning conditions to ensure that the technology is suitable also for home use and not only in standardized environments.

YOLOv8 provides several different pre-trained models for pose estimation with different levels of accuracy and computational cost. These different versions should also be more rigorously tested for finding a suitable balance between usability and accuracy.

ACKNOWLEDGEMENTS

This work was supported by the "Fonden för teknisk undervisning & forskning" (TUF) under Grant 1580, a fund for supporting education and technical research at Arcada University of Applied Sciences, Finland.

REFERENCES

[1] T. Hellstén, J. Arokoski, T. Sjögren, A. M. Jäppinen and J. Kettunen, The Current State of Remote Physiotherapy in Finland: Cross-sectional Web-Based Questionnaire Study.

JMIR Rehabil Assist Technol, vol. 9, pp. e35569, Jun. 2022, doi:10.2196/35569

[2] P. Truter, T. Russell and R. Fary, The Validity of Physical Therapy Assessment of Low Back Pain via Telerehabilitation in a Clinical Setting. Telemedicine journal and e-health, vol. 20, pp. 161-167, Feb. 2014, doi:10.1089/tmj.2013.0088

[3] R. B. Burns et al., Using telerehabilitation to support assistive technology. Assist Technol, vol. 10, pp. 126-133, Oct. 1998, doi: 10.1080/10400435.1998.10131970

[4] M. Capecci et al., A Hidden Semi-Markov Model based approach for rehabilitation exercise assessment. Journal of Biomedical Informatics; J Biomed Inform, vol. 78, pp. 1-11, Feb. 2018, doi: 10.1016/j.jbi.2017.12.012

[5] L. F. Pastana Ramos, T. C. S. Vilacorta-Pereira, J. D. S. Duarte, E. S. Yamada and B. L. Santos-Lobato, Feasibility and effectiveness of a remote individual rehabilitation program for people with Parkinson's disease living in the Brazilian Amazon: a randomized clinical trial. Front Neurol, vol. 14, pp. 1244661, Aug. 2023, doi: 10.3389/fneur.2023.1244661

[6] H. Spindler, K. Leerskov, K. Joensson, G. Nielsen, J. J. Andreasen, and B. Dinesen, Conventional Rehabilitation Therapy Versus Telerehabilitation in Cardiac Patients: A Comparison of Motivation, Psychological Distress, and Quality of Life. Int J Environ Res Public Health, vol. 16, pp. 1-16, Feb. 2019, doi: 10.3390/ijerph16030512

[7] F. Fatoye et al., The Clinical and Cost-Effectiveness of Telerehabilitation for People With Nonspecific Chronic Low Back Pain: Randomized Controlled Trial. JMIR Mhealth Uhealth, vol. 8, pp. e15375, Jun. 2020, doi: 10.2196/15375

[8] A. Rintala et al., Effectiveness of Technology-Based Distance Physical Rehabilitation Interventions for Improving Physical Functioning in Stroke: A Systematic Review and Meta-analysis of Randomized Controlled Trials. Arch Phys Med Rehabil, vol. 100, pp. 1339-1358, Jul. 2019, doi: 10.1016/j.apmr.2018.11.007

[9] M. Nelson, M. Bourke, K. Crossley and T. Russell, Telerehabilitation is non-inferior to usual care following total hip replacement—a randomized controlled non-inferiority trial. Physiotherapy, vol. 107, pp. 19-27, Jun. 2020, doi: 10.1016/j.physio.2019.06.006

[10] R. Del Pino et al., Costs and effects of telerehabilitation in neurological and cardiological diseases: A systematic review. Frontiers in Medicine, vol. 9, pp. 1-14 , Nov. 2022, doi: 10.3389/fmed.2022.832229

[11] B. Debnath, M. O'brien, M. Yamaguchi and A. Behera, A review of computer vision-based approaches for physical rehabilitation and assessment. Multimedia Systems, vol. 28, pp. 209-239, Jun. 2022, doi: 10.1007/s00530-021-00815-4

[12] N. Goldfarb, A. Lewis, A. Tacescu and G. S. Fischer, Open source Vicon Toolkit for motion capture and Gait Analysis. Comput Methods Programs Biomed, vol. 212, 106414, Nov. 2021, doi: 10.1016/j.cmpb.2021.106414

[13] T. Hellstén, J. Karlsson, M. Shamsuzzaman and G. Pulkkis, The potential of computer vision-based marker-less human motion analysis for rehabilitation. Rhabilitation process and outcome, vol. 10, Jul. 2021, doi: 10.1177/11795727211022330.

[14] E. Hannink et al., Validity and feasibility of remote measurement systems for functional movement and posture assessments in people with axial spondylarthritis. Healthcare Technology Letters, vol. 9, pp. 110-118, Dec. 2022, doi: 10.1049/htl2.12038.

[15] G. Jocher and J. Qiu, Ultralytics YOLO (Version 8.0.0) [Retrieved: 3.4.2023]. Available from: https://github.com/ultralytics/ultralytics.

[16] T-Y. Lin et al., " Microsoft coco: Common objects in context", Computer Vision–ECCV 2014: 13th European Conference, Zurich, Switzerland, Springer, Sept. 2014, pp. 740-755.

[17] J. Cibere et al., Reliability of the hip examination in osteoarthritis: effect of standardization. Arthritis Rheum, vol. 59, pp. 373-381, Feb. 2008, doi: 10.1002/art.23310

[18] N. B. Reese and W. D. Bandy, Joint Range of Motion and Muscle Length Testing, 2nd ed. Missouri: Elsevier, pp. 59-65, 2009.

[19] L. Brosseau et al., Intratester and intertester reliability and criterion validity of the parallelogram and universal goniometers for active knee flexion in healthy subjects. Physiotherapy Research International, vol. 2, pp. 150-166, Mar. 1997, doi: 10.1002/pri.97.

A Case Study for Scoliosis: How MLOps Can Help Reduce AI Challenges in Health Care?

Gábor György Gulyás
Vitarex Stúdió Ltd
Budapest, Hungary
gabor@gulyas.info

Janis Lapins
Data Science
Spicetech Gmbh
Stuttgart, Germany
janis.lapins@spicetech.de

Attila Csaba Kiss
Vitarex Stúdió Ltd
Budapest, Hungary
kiss.csaba@vitarex.hu

Abstract—**The integration of Artificial Intelligence (AI) into healthcare diagnostics represents a significant advancement, particularly in the screening for conditions, such as scoliosis. This paper discusses the development, implementation, and evaluation of the Posture Buddy (PB) device, a machine vision driven tool designed to enhance the efficiency of scoliosis screening among school-aged children within the Hungarian health visitor system. Through the lens of Machine Learning Operations (MLOps) practices, our case study demonstrates the pivotal role of MLOps in overcoming operational hurdles at the intersection of eHealth and AI. The field-testing of PB revealed that within the context of low light conditions and slight side viewing angles the device performance decreases. In a later phase of the project, the pose estimation model of the device was put through model validation, observing the same flaw. Through these findings, the importance of proactive validation of AI models in healthcare is highlighted, whereas it also underscores the need to use MLOps to enable continuous deployment through the lifecycle of ML-based medical tools.**

Index Terms—**machine learning, edgeML, health care, MLOps, pose estimation.**

I. Introduction

The integration of Artificial Intelligence (AI) in healthcare has opened new frontiers in diagnostics, treatment planning, and patient care, offering the potential to significantly enhance the accuracy and efficiency of medical services. However, the adoption of AI technologies in healthcare is fraught with challenges such as the reliability, validation, and operationalization of AI models. Scoliosis, a condition characterized by an abnormal lateral curvature of the spine, affects millions worldwide, and traditional screening methods, while effective, are labor-intensive. This underscores the need for automated AI-driven solutions that could improve the speed of screening and also allow better documentation of results. Eventually, such solutions could become more accurate than traditional screening.

This paper presents a case study focused on the development of a Machine Learning (ML) based device for scoliosis screening, highlighting the pivotal role of Machine Learning Operations (MLOps) practices in overcoming the prevalent challenges. The study is anchored in the development and field-testing of Posture Buddy (PB), a device aimed at enhancing the screening process for spinal abnormalities among school-aged children. While machine learning, and machine vision in particular, have a whole host of potential uses in health care, it is crucial that malfunctions are handled, or avoided in advance.

During the field trial of PB, it emerged that its performance decreased under poor lighting conditions and when looking at the patient from aside. Had this been identified beforehand, the users could have been notified in advance, or even deployed an updated model version to mitigate this issue. However, it was not identified earlier, but at an advanced phase, a virtual validation tool called VALICY enabled us to identify these same shortcomings in the model (this validation was an international collaboration in the IML4E [Industrial Grade Machine Learning] project [6], for which PB served as a testing ground). This experience underscores the importance of thorough validation before the first deployment, and the utilization of MLOps infrastructure, to preemptively correct such errors before they impact users.

The paper is structured as follows. In Section II, the Hungarian health visitor system is described to the point of outlining their work in scoliosis screening. Following that, in Section III, PB is presented, a device that was developed to help and enhance the screening process. Its details considering hardware, software and machine learning, and how it was evaluated in our field study are described. In Section IV, the MLOps pipeline is presented, along with model evaluation methods. Section V provides validation details with a black-box validation tool, and the current work is concluded in Section VI.

II. Health Visitors and Scoliosis Screening

Health visitors play a crucial role in public health in Hungary. They provide essential services and support to individuals and families, particularly in the realm of preventive healthcare and early intervention, with a focus on the health of children.

A. Their History and Current Service

The health visitor service was established in 1915 under the name Stefánia Association, with its main goal being the protection of mothers and infants [21], [22]. The Green Cross Health Service started in 1927, whose health visitor service operated from 1930 to 1944, with its scope of activities extending to school-age groups. From then on, health visitors were systematically involved in school health care in educational institutions. After World War II, the two health visitor services merged, and their current work is based on The Law on Public Education (1993). This provides ground for students to have the right to receive regular health supervision and care.

There are two types of health visitors: school and district health visitors. School health visitors primarily focus on providing healthcare services within educational institutions. On the other hand, district health visitors operate within communities, offering healthcare services directly to families, particularly focusing on maternal and child health, preventive care, and health education. Both play important roles in public health, with school health visitors emphasizing school-based interventions and district health visitors focusing on community-based health promotion and support.

B. *Posture Screening in Schools*

Health visitor screening examinations are conducted for children aged 3-18 years, where they collaborate in the provision of school health tasks [22]. By law, school health care contains tasks to be performed independently by the health visitor, including screening examinations for specific age groups, such as assessing height, weight, physical development, identifying psychological, motor, mental, and social development and behavioral problems, among others. In addition, it is important to document the tasks performed.

All these circumstances enable the testing of planned tools with the involvement of health visitors. Our project targeted musculoskeletal screening examinations, with a particular focus on spinal disorders. Considering that screening examinations are carried out by multiple professionals in various age groups every two years, the introduction of a digital measurement application could enhance documentation and comparison of results.

C. *Scoliosis Screening Protocol Followed by Health Visitors*

The procedure for musculoskeletal examination and the positions and movements to be considered during the examination setup are outlined based on the guidelines provided in the school health manual edited by Dr. Anna Aszmann [1]:

1) The student stands with their back to the examiner. They wear only underwear on their upper body. The examiner observes the student's posture, focusing primarily on the trunk-arm triangle, shoulder deviations, and spinal curvatures.
2) The examiner asks the student to raise both arms towards the ceiling. They observe whether the student can compensate for any observed abnormalities with the back muscles.
3) The student leans forward with extended arms towards the floor. Here, the health visitor assesses the strength of the back muscles and the ability to compensate for abnormalities.
4) The examiner instructs the student to straighten up and lower their arms to their sides loosely, then to turn to one side (with the legs as well). With closed eyes, the student raises both arms horizontally and holds them for about 30 seconds. Any changes in posture during this test, such as leaning back or tilting the pelvis forward, may indicate postural problems or weakness.
5) The student lowers both arms to their sides and faces the examiner, allowing any potential pelvic abnormalities and chest deformities to be observed.

During this protocol, health visitors are looking for postural disorders and signs of scoliosis.

III. THE *Posture Buddy* DEVICE

The development of this device was done within the IML4E project as a use case of the project [6] (2021-2024). The purpose behind developing and validating the PB tool is to enhance preventive screenings for musculoskeletal abnormalities (such as spinal and other postural disorders) among students, moving beyond the current method of assessment based on visual inspection, lacking visual documentation and an objective basis for comparison.

A. *Hardware and Appearance*

Initially, a decision on the platform was needed. Developing a mobile application seemed like a natural solution, since the majority of potential users already own one. However, the variety of OS (Operating System) versions, lacking sufficient support to properly run the deep learning based apps, posed a challenge. Additionally, at the project's inception, except for the top-tier mobile phones (which were not widespread in Hungary), our algorithms ran slowly on mobile devices, typically less than one frame per second.

Since health visitors in rural areas have unstable network connection, the posture analysis program needed to function without a network connection. Moreover, for privacy reasons, it was desirable for the evaluation to be performed directly on the device (edge ML); making it impossible to move the evaluation behind a secured API (Application Programming Interface) in the cloud. These constraints necessitated storing and running the machine learning model on the device itself. However, running larger models or updating models on mobile devices is rather difficult, thus prompting exploration of alternative options.

Alternatively, the application could be developed for single-board computers, such as the Raspberry Pi [8]. Such devices are small-sized, low-power, relatively inexpensive, and possess much of the functionality of a traditional computer. Therefore, they offer much more flexibility, but also bring in additional challenges (e.g., providing a proper housing, screen and peripherials). Besides the widely known Raspberry Pi, the Nvidia Jetson Nano system [7] seemed the most suitable for the task, as it features GPU-like (Graphics Processing Unit) hardware acceleration for machine learning (with 128 CUDA [Compute Unified Device Architecture] cores). But, as Nvidia Jetsons were globally unavailable in 2021, we opted for the Raspberry Pi 4.

This all resulted in a uniquely designed, compact-sized computer with its own custom 3D-printed housing developed

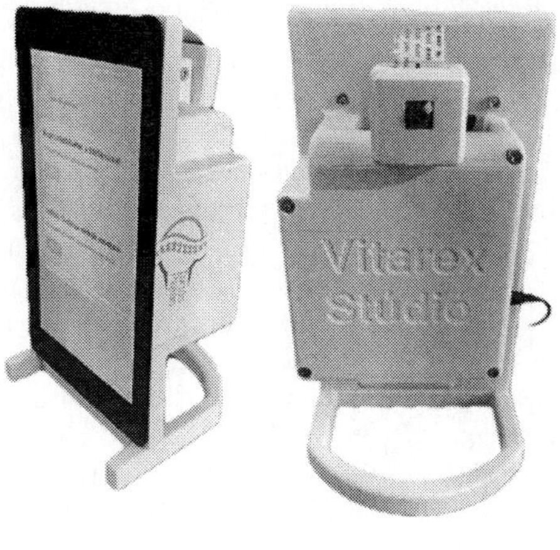

Fig. 1: Two photos of the Posture Buddy device.

by Vitarex Stúdió Ltd, depicted in Figure 1. Its main board is a Raspberry Pi 4 (4GB or 8GB), with an appropriate camera, a 7-inch touchscreen, and a USB-C type charger. The housing is designed to leave the Raspberry Pi ports accessible, allowing peripherals, such as external screens, keyboards, and mice to be connected, enabling the saving of results to a USB drive if needed. When turned on, the posture analysis program starts automatically.

B. Software

The posture analysis is a full screen application that loads automatically after turning the device on. It follows a five-step examination procedure based on the spinal examination protocol described in Section II-C. During the examination, students need to position themselves in front of the camera according to specific instructions displayed on the screen, then the health visitor captures the image. The software automatically detects and displays key points of the human figure in the images (when possible), allowing manual correction if needed. Throughout the five steps, the system continuously calculates and records data indicating the extent of spinal curvature deviation. Upon completion of the examination, it displays all captured images and analytical results.

The software architecture is designed to leverage the full potential of the open computing platform: PB is a lightweight web application (running fully locally). This enables an integrated client-server architecture within a single device, it is easy to change and to update, even remotely. The client side uses web technologies, such as HTML (HyperText Markup Language), CSS (Cascading Style Sheets), and JavaScript, and the server side is a Flask Python app [10], running machine vision based on OpenCV [11] and other libraries.

The server-side has a modular structure, separating different functional units to promote maintainability and clarity of the codebase. For example, the data submission module (for MLOps) is responsible for anonymizing images generated during examination. The metrics module performs calculations related to individual examination steps. Another module handles the integration with the Stefánia Registration System [12] for health visitors (which is used by the majority of health visitors in Hungary), enabling the recording of examination images within the system. Another module is responsible for exporting examination summaries to a USB drive.

In order to deliver fixes to errors, a software update mechanism was implemented, enabling the distribution of new software versions. The new software version is uploaded as a Github Release Asset and the software periodically checks if a new version is available. New versions are downloaded with dependencies attached, and are installed. Finally, the device is restarted. This mechanism ensures the PB device remains functional and up-to-date, enhancing user experience and device reliability.

C. Machine Vision Algorithms

The software solution utilizes multiple machine learning models. The first model determines the coordinates of key points of the human figure in the images (if possible), then additional models serve to separate students from the background and separate the shape and contour of the student's figure (if needed).

1) Pose Estimation: Before entering into the development, models are compared based on their accuracy (by considering their MSE; Mean Square Error) and their processing speed in terms of FPS (Frame Per Sec). At the time, the most widespread model was PoseNet [2]. The output stride (with possible values of 8, 16, 32), regulates the model's processing accuracy, where higher values result in faster but less precise processing. The model also has a multiplier parameter (with possible values of 1.01, 1.0, 0.75, and 0.50) that controls the depth of convolutional processing, where higher values offer more precision but slower processing. Two settings were examined: a rapid but less accurate setup with output stride = 16 and multiplier = 0.5, and a slower yet more precise configuration with output stride = 32 and multiplier = 1.01. OpenPose [3] offered multiple models, the one with 25 keypoints were selected, as that proved to be the most efficient. Due to video memory constraints, it was tested with resolutions of 128x128 and 160x160. For the TRT_Pose system [4], the ResNet-18 model (18 layers deep) was used as it yielded better results based on our preliminary measurements.

In the measurements, it was observed that the TRT_Pose model delivered the best performance by simultaneously achieving the highest image processing speed while producing the fewest errors (5.6 FPS, 13 MSE). The OpenPose models also performed well with minimal errors, albeit at significantly lower image processing speeds (3,6 FPS with MSE around 17). PoseNet lagged behind in both evaluation criteria compared to these results (2.1 FPS and MSE 78; 3.7 FPS and MSE 108).

In each step of the pose estimation process, after taking the picture, PB allows health visitors to improve the results. That is, a screen is loaded where all detected keypoints are displayed over the photo. The health visitor is then allowed to move these points for correction. In the case of a correction, the provided information could be used for improving the model performance.

2) Machine Vision Algorithms: As described in Section II-C, five different metrics have been used for each step in the protocol. All photos are taken in a standardized setting, allowing asymmetries to emerge.

Standing backwards. First, the keypoints are calculated with TRT_Pose, then check the angle defined by the keypoints for the hips and shoulders (horizontal asymmetry). Greater angles reported by this metric can mean a more serious case of scoliosis. (cf. Figure 2a)

Standing backwards, hands raised. In this step, the keypoints are also calculated, and then measure vertical asymmetries between shoulder and hip points. Deviations are measured by the visible tilt of these bodypoints. (cf. Figure 2b)

Leaning forwards. In this step, the student bends forward with their arms ahead, leaving insufficient visibility for keypoint-based measurements. Instead, the silhouette of the back is determined, and its inclination angle is calculated by polygon fitting. The larger the angle, the higher the chance of more advanced scoliosis.

Turned to side, hands raised in front. When they are turned to their side, keypoints can be used again. In this case, keypoints are used combined with polynomial silhouette mapping to determine how bent the spine is.

Standing in front of the camera. This step uses keypoints to determine the vertical asymmetry of keypoint pairs on each side: it gives an asymmetry score about how the distances of keypoints are proportional to each other.

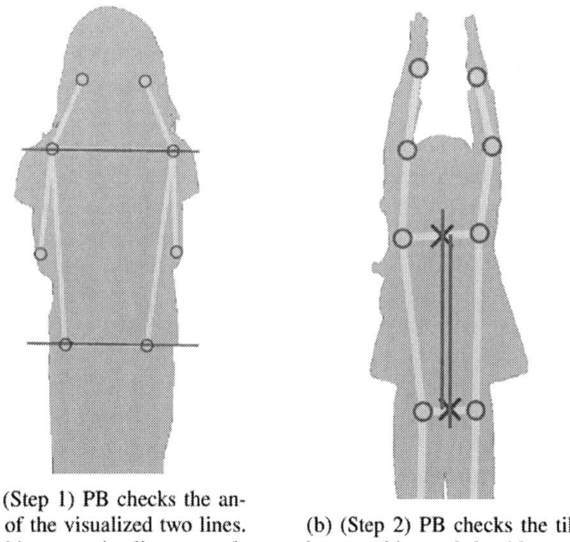

(a) (Step 1) PB checks the angle of the visualized two lines. In this case, the lines are almost in parallel.

(b) (Step 2) PB checks the tilt between hips and shoulders, as displayed by the lines.

Fig. 2: Visualization examples of pose evaluation, using silhouettes only for preserving privacy.

D. Evaluation of the Posture Buddy in a Field Study

Prerequisites for participating in the program included informing their employers, to provide them detailed information about the program, covering its objectives, content, and the roles of health visitors, parents, and students. Participating students were provided detailed information and consent forms (to their parents), who were also verbally informed about the details of the examination, its purpose, data protection measures, and the anonymous handling of data. The trial itself took place in official premises of the institution, involving students who volunteered and had consent forms, outside of regular class hours.

Overall, it can be said that the health visitors conducting the tests provided a realistic picture and assessment of the technical and professional usability of the pose estimation device. Their thorough and detailed technical evaluation was supported by quantified data. Their observations included that it was difficult to set up the internet on the device (for Eduroam), and it would be beneficial if the height of the device could be adjusted like a camera tripod.

They agreed, that after having some features corrected and adding further minor refinements, PB should be a useful tool for school health visitor work. Due to its digital nature, health visitors recommend that besides uploading the resulting examination data set to health visitor programs (such as Stefánia), it should also be uploaded to the EESZT (Electronic Health Service Space) for further use by general practitioners, pediatricians, school physicians, and orthopedic specialists. They found it suitable to attach the findings of the basic screenings, supported by measured data, to the health visitor reports, and to substantiate and justify the referrals for further medical and specialist examinations.

Regarding the operation of the device, one of the health visitors made a significant observation: under poor lighting conditions, the device provides a somewhat inaccurate predictions when viewed from an angle. This health visitor worked in two schools, and in one, their room was long, with poor lighting, and the power cable was too short to provide a good view over the examined student (just from the side). This led to a notably higher error rate of the pose estimation model.

Considering all their comments, a second round of testing is currently run at the time of the writing of this paper. This second round of testing is country wide, involving more schools and health visitors. A mobile application version of the PB device is also developed.

IV. Machine Learning Operations

The former and similar situations where malfunctions have occurred could have been detected in advance with the appropriate validation tools. Even if the devices were already deployed, health visitors' attention could have been drawn to this issue. This is where MLOps, an emerging field can help. MLOps tools cover a wide range, including tools related to failure detection of models, transfer learning and distribution of new model versions, among many others.

A. Our MLOps Infrastructure and Updates

One of the main objectives of the project was to create an MLOps infrastructure that is integrated with the PB.

1. Pipeline. Machine Learning Operations is a comprehensive approach to systematically and efficiently manage the lifecycle of machine learning models [14]. This lifecycle includes the continuous training, evaluation, deployment and monitoring of models. In the context of our project this meant that an MLOps system was created to continuously train, monitor and improve the keypoint detection model built in the PB software.

The infrastructure was mainly developed in Python, which is suitable for data science and machine learning tasks. The chosen deep learning framework was Pytorch [15] because of its popularity and ease of use. The chosen base keypoint detection model was TRT_Pose [4] as previously explained. A pivotal component of the infrastructure is the model training pipeline. It serves the purpose of training keypoint detection models with different hyperparameters and settings.

These settings include:

- the datasets the model is trained and evaluated on
- the model architecture
- the starting weights the model uses
- the ratio between the training and testing sets
- the shapes of the model input and output
- the transformations that are performed on the input images
- type of the optimizer
- the learning rate
- the number of epochs the training lasts

The whole pipeline can be controlled by a configuration file, which contains the values of the pipeline parameters.

The pipeline is structured into five sequential phases. Ingest is responsible for loading and preprocessing datasets, consisting of images and annotations, into a usable format. The next phase is split, where the dataset is sub-divided into distinct training and testing subsets. This is followed by transform, where random transformations to the training data were applied, which may include adjustments in rotation, scale, translation, and color modifications. In the training phase, the initialized model was trained using the specified parameters on the training dataset. Finally, in the evaluation phase a customized evaluation to measure the model's accuracy in keypoint detection was performed. The specifics of this evaluation are explained in a later paragraph. The pipeline is executable both as a standalone Python script and within an interactive Jupyter Notebook environment, offering flexibility in parameter adjustments and modular execution of steps.

2. Evaluation. To accurately assess the efficacy of the keypoint detection models trained via the outlined pipeline, the cocoapi [5] and coco-analyze [16] libraries were used. In the evaluation step the trained model is used to make predictions on the validation images and these predictions are compared to and analyzed with the ground truth values. Through this analysis the Average Precision (AP) and Average Recall (AR) concerning keypoints was calculated.

The coco-analyze library defines a number of error types and calculates the potential improvement in the average precision and average recall metrics if these errors were corrected. Some of these build on the term Object Keypoint Similarity (OKS), which is a metric used to evaluate the accuracy of detected keypoints. It calculates the similarity between the predicted keypoints and the ground truth, considering the distance between them, the standard deviation of the keypoints and the scale of the object. The defined errors are the following:

- **Miss:** the miss score means that the detected keypoint is not close to any body part.
- **Swap:** in this case, the detection close to a body part of a different person.
- **Inversion:** the keypoint is matched to another body part of the same person (e.g., mismatching keypoints of legs or hands).
- **Jitter:** correct keypoint identification, location slightly differs.
- **Score:** close to a ground truth annotation, when two detections are identified, it is the detection with the highest level of confidence that ends up having the lower OKS score.
- **Background false positive:** detections without a ground truth annotation match.
- **False negative:** missed detections.

These metrics are graphically represented, and a detailed report is generated. This facilitates a straightforward comparison of model performances across various validation datasets highlighting the model's strengths and areas for improvement.

3. Mlflow. Mlflow [9] is a popular library that can be used to streamline machine learning development, including tracking experiments (both parameters and results), packaging code into reproducible runs, and sharing and deploying models. The developed model training pipeline is integrated with Mlflow library to utilize its components. These allow the straightforward comparison of various pipeline executions in the Mlflow user interface. The Model Registry component of Mlflow provides a repository for storing the models that have been trained in the pipeline, presenting them in an easily deployable format.

4. Deployment. In the Mlflow user interface all of the available models can be easily compared based on their parameters and metrics. If the model with the best results is chosen, it can be released with running a script that uploads the model to an easily accessible cloud storage service, which in our case is a Github Release page. The trained models are stored and deployed in TorchScript format, in which the models can be packaged without the need to define their original architecture in the production environment. Also, these can be used in non-Python environments and can be optimized for use on edge devices. PB checks if a new model version is available on every start up, and runs an update script if there is any.

5. Monitoring. PB users, at the end of the examination process, can choose to send the original and the corrected keypoint detections to the server of Vitarex along with the anonymized images (faces removed). This way the collected data can be used to evaluate the performance of the model in production as well as to create new datasets to improve the model.

V. Model Validation with VALICY

A. How VALICY Works

VALICY is an AI-based black box testing environment, which allows the virtual validation of multi-dimensional AI based classification systems and complex software, developed by Spicetech GmbH since 2017 [17]. It creates test proposals for the black box application under test which in turn are evaluated and feed back to VALICY.

Through evolution and a competing AI swarm, awareness of the problem's nature increases as more evaluation points provide additional training data. This process helps identify safe or unsafe operational areas. Testing ceases once a predetermined number of runs conclude and either a residual uncertainty is quantified or the desired certainty level is achieved, signaling the end of test proposals.

VALICY's components are:

- a Python framework within a Docker container,
- a server with a set of CPUs,
- a frontend in the browser to display all job results along with the possibility to do cluster analysis of the results and get corresponding characteristics,
- a REST-API to exchange data securely [20],
- a Grafana dashboard to monitor operation during runs, and
- a database.

The Python framework operates across three layers: a Docker layer hosting multiple job instances on a single server, a job instance layer managing workload and enabling the creation of new AI instances (i.e., pre-configured models), and an AI instance worker layer. Depending on the nature of the response of the application under test (fast vs. slow), a different number of AI models run in parallel to generate proposals. The default value of competing AI models is three.

In this framework, grid points define the range of input parameters, serving as the initial dataset for AI to generate decision-making proposals. These proposals seek to pinpoint the boundary where outcomes shift from True to False, with their precision improving through iterative evaluations and feedback. AI models log each run's outcomes and settings, leveraging this history to enhance future predictions and decision-making processes.

Testing a "black box" application requires only the input parameters—name, range, type (continuous or discrete)—and target parameters—name, threshold, direction (upper, lower), and desired certainty. The process begins by sampling these boundaries and submitting them to the black box via API, with responses stored in the VALICY database and used to inform AI models.

To avoid overlooking key areas, the system intersperses AI-driven evaluations with randomly generated points. Following initial sampling, AI analyses feedback from test points to refine its models, subsequently proposing and assessing potential evaluations based on their likelihood of success. This iterative refinement, informed by direct comparison between predicted outcomes and actual black box feedback, progressively enhances the AI model abilities to accurately predict near the decision boundaries, thus improving its effectiveness in identifying viable configurations for future tests.

To evaluate the coverage and global certainty of the test space for VALICY's stopping criterion, a "geometrical" instance identifies and fills the largest unsampled volumes by placing test points at positions furthest from previously sampled points.

VALICY halts a job based on two criteria: achieving the predetermined certainty level consistently after a set number of runs, or when all configured test points have been used. Throughout the job, results are continuously sent to the application. Comprehensive analyses including performance comparisons of AI models, clustering of results for True and False values, identification of points closest to cluster centers as representative, and outlier detection through various methods (e.g., neighbor distance, cluster center distance) are available for export via the frontend. Plans are underway to define a "safety envelope", encompassing volumes of True values at a certain distance from the decision boundary, to ensure reliable operation within specified parameter ranges under the defined certainty level.

B. Model Evaluation with Blender

Blender [23] is an open-source 3D computer graphics software used for creating animated films, visual effects, 3D-printed models, motion graphics. Blender was used to create a basic scene, which tries to simulate the real-world use of PB. It consists of a realistic 3D human model, a pale wall as background, a camera and a light source to mimic real-world conditions. The camera is positioned to capture the human model, this way images of the human can be generated.

To run virtual validation with VALICY, Vitarex modified the publically accessible gitlab repository [13] and adjusted the provided validation sample Python code to account for the input parameter variations. The 4 input parameters of the validation process are lighting, radius (distance), phi (angle in the horizontal plane, with 0/360° being a frontal), theta (angle with respect to the horizontal plane).

During the testing phase every time a set of parameter values is received from VALICY, a Blender Python script is executed, which sets the position of the camera and the intensity of the light source. After that an image is captured of the human model. The image is fed to the keypoint detection model. Then resulting detections are evaluated primarily by counting the number of detected keypoints. The prediction is considered correct if at least 80% of the visible keypoints are detected. The prediction outcome is relayed back to VALICY, prompting

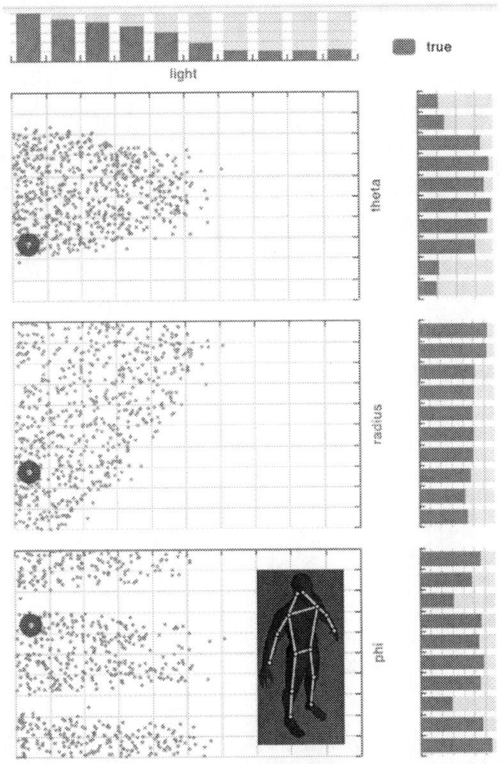

Fig. 3: Distribution of True results for the virtual validation runs of the key point detector plotted over the input parameter combinations (red circle is corresponding to the example).

the proposal of new parameter sets. For subsequent validation efforts, this threshold may be raised to enhance robustness.

After the completion of the testing, its results are displayed on the VALICY dashboard. See evaluation details in Figure 3. Based on the results, it can be concluded that the stronger the light source is, the bigger the distance between the camera and the person should be. It can also be clearly seen that the model cannot detect correctly when the phi parameter is between 77-110 degrees and 252-292 degrees, meaning when camera faces the side of the person. It is the same situation when the theta parameter is above 150 or below 33 degrees, which means the camera should not be placed too much above or below the person. In conclusion, one can say that, with the help of VALICY validation, one could determine the exact limits of the keypoint detection model, even more precisely than with the health visitors' findings.

VI. Conclusion

The development and field-testing of the PB device for scoliosis screening within the Hungarian health visitor system showcases the essential role of Machine Learning Operations (MLOps) in the successful deployment of AI technologies in healthcare. In this paper, the healthcare ecosystem in Hungary was shown, which is responsible for the screening of kids in schools. The development of an edgeML device (PB) for scoliosis screening was presented, and its potential was discussed that was tried out in a field study. It turned out that PB had some flaws that were rooted in the misbehaviour of the used ML model. These flaws could have been identified before its first real life deployment if the validation would have been done at an earlier stage (the validation was an international collaboration in the IML4E project [6], for which the device served as a testing ground).

Acknowledgment

The authors would like to acknowledge the contribution of their colleagues and partners to the work described in this paper. Our thanks go to former members of the development team at Vitarex, in particular to Dániel Kuknyó, Tamás Csarnó and Bence Várhidi, who all contributed to the development of PB. The authors would like to express their gratitude to Marianna Várfalvi for her professional advice (Health Visitor, chair of MVSZSZ, Professional Association of Hungarian Health Visitors).

Project no. 2019-2.1.1-EUREKA-2020-00016 has been implemented with the support provided by the Ministry of Innovation and Technology of Hungary from the National Research, Development and Innovation Fund, financed under the 2019-2.1.1-EUREKA funding scheme.

References

[1] A. Aszmann et al., "Health Protection in Public Education". Antikvárium Kiadó (Antiquarian Publisher), 2005.
[2] PoseNet. https://blog.tensorflow.org/2018/05/real-time-human-pose-estimation-in.html, accessed on 2024-04-26.
[3] Openpose. https://github.com/CMU-Perceptual-Computing-Lab/openpose, accessed on 2024-04-26.
[4] TRT_Pose. https://github.com/NVIDIA-AI-IOT/trt_pose, accessed on 2024-04-26.
[5] Coco API. https://github.com/cocodataset/cocoapi, accessed on 2024-04-26.
[6] The IML4E project. https://iml4e.org, accessed on 2024-04-26.
[7] NVIDIA Jetson Nano. https://developer.nvidia.com/embedded/jetson-nano-developer-kit, accessed on 2024-04-26.
[8] Raspberry Pi 4 model B. https://www.raspberrypi.com/products/raspberry-pi-4-model-b/, accessed on 2024-04-26.
[9] Mlflow: A machine learning lifecycle platform. https://github.com/mlflow/mlflow, accessed on 2024-04-26.
[10] Flask Web Framework. https://flask.palletsprojects.com/en/3.0.x/, accessed on 2024-04-26.
[11] OpenCV. https://opencv.org, accessed on 2024-04-26.
[12] Stefánia Registration System. https://vitarex.hu/Stefania, accessed on 2024-04-26.
[13] VALICY repository. https://github.com/SpicetechGmbH/Valicy-Interface-Example, accessed on 2024-04-26.
[14] Google Cloud Architecture Center. Mlops: Continuous delivery and automation pipelines in machine learning. https://cloud.google.com/architecture/mlops-continuous-delivery-and-automation-pipelines-in-machine-learning, accessed on 2024-05-22.
[15] A. Paszke et al. Pytorch: An imperative style, high-performance deep learning library. In *Advances in Neural Information Processing Systems 32*, pp. 8024-8035, Curran Associates, Inc., 2019.
[16] M.-R. Ronchi and P. Perona. Benchmarking and error diagnosis in multi-instance pose estimation. In *The IEEE International Conference on Computer Vision (ICCV)*, Oct 2017.
[17] Fortissimo908-success-story. https://www.fortissimo-project.eu/en/success-stories/908/massively-parallel-virtual-testing-of-safetyrelevant-driving-systems, accessed on 2024-04-26.
[18] J. Lapins et al., Massively Parallel Virtual Testing of Safety-Relevant Driving Systems. *Proceedings of 7th AutoTest Technical Conference*, Stuttgart, 2018.

[19] K. Gao, S. Hekeler, and M. Kütemeyer, Combination of virtual and real live tests of safety relevant driving functions (Original title: Kombination von virtuellen und realen Tests sicherheitskritischer Fahrfunktionen). ATZ Elektron 15, pp. 66–70 (2020). https://doi.org/10.1007/s35658-020-0268-1, accessed on 2024-04-26.
[20] VALICY REST-API description. https://api.valicy.de/docs, accessed on 2024-04-26.
[21] B. Pukánszky and A. Németh (1996): Neveléstörténet (Education history), Nemzeti Tankönyvkiadó Rt (National Textbook Publishing Co).
[22] M. Kachlichné Dr. Simon and M. Várfalvi (2020): Health visitors' web history museum. http://mvszsz.hu/index.php/hu/webmuzeum, accessed on 2024-05-22.
[23] Blender website. https://www.blender.org/, accessed on 2024-05-22.

Leveraging Voice for Early Detection of Chronic Kidney Disease: Enabling Continuous Monitoring in Remote Healthcare

Kangbeen Ko
Gwangju Inst. of Science and Technology
Gwangju, Republic of Korea
eyeoftyphoon@gm.gist.ac.kr
0009-0006-5286-9842

Jiwon Ryu
Seoul National University Bundang Hosp.
Seongnam, Republic of Korea
bboddo5@hanmail.net
0000-0002-2372-8948

Sejoong Kim
Seoul National University Bundang Hosp.
Seongnam, Republic of Korea
sejoong2@snu.ac.kr
0000-0002-7238-9962

Abstract—**Chronic Kidney Disease (CKD) represents a globally prevalent condition characterized by the gradual loss of renal function over time. The covert progression of CKD accentuates the necessity for regular and continuous inspection. Conventional diagnostic methods for CKD, including blood and urine analyses to estimate the Glomerular Filtration Rate (GFR) and to measure the urine Albumin-Creatinine Ratio (uACR), while effective, are invasive and often fail to facilitate early detection due to the asymptomatic progression of CKD in its initial stages. To tackle these limitations, we propose a novel, non-invasive diagnostic technique to enhance the early detection and management of CKD. This technique utilizes the patients' voice features, caused by respiratory muscle weakness and vocal chord swelling in patients with CKD, as an auxiliary indicator, leveraging machine learning algorithms to identify subtle changes in voice patterns that may correlate with CKD progression. Our method demonstrated a diagnostic accuracy of 0.86, quantified by the F1 score, and showed promising potential as a supplementary diagnostic tool. Implementing this technique paves the way for its integration into telemedicine platforms, offering a promising avenue for remote monitoring and managing CKD patients. This breakthrough advances our understanding and capability in the early diagnosis of CKD. It expands the potential for remote healthcare delivery, ensuring timely intervention and improving patient outcomes in managing kidney conditions.**

Keywords—chronic kidney disease; automatic classification; machine learning; explainable artificial intelligence.

I. Introduction

Kidney disease is a decrease in the Glomerular Filtration Rate (GFR), which refers to the degree of waste removal in the kidney, or when the signs of structural or functional decrease of the kidney are detected by blood, urine, radiation, or other kidney pathology tests [5]. It affects approximately 8% to 16% of the world's population and is the leading cause of death worldwide [19]. Among them, a case where this decrease in kidney function lasts for more than three months is called Chronic Kidney Disease (CKD). This status can cause not only kidney failure but also some other adverse outcomes such as cardiovascular disease and ultimately cause the need for dialysis or renal replacement therapy [16].

The clinical research underscores the importance of prompt identification and therapeutic intervention for stage 3 CKD to avert additional deterioration of renal function and the advancement to renal failure [14]. The study presents compelling evidence that early detection, accurate staging, and suitable management of CKD can mitigate these negative consequences and diminish the overall impact of the disease. However, one of the key challenges in the early detection of CKD is its silent progression. In its early stages, symptoms are often absent [17], and traditional diagnostic methods such as blood and urine tests require regular medical check-ups, which hinders early detection and continuous monitoring. Accordingly, there is a growing demand for auxiliary indicators that can diagnose CKD non-invasive and iteratively.

To find them, we focused on the fact that CKD impacts multiple body systems, such as the cardiovascular, nervous, musculoskeletal, immune, endocrine, metabolic, and respiratory systems. Among them, the close functional relationship between the lungs and kidneys in maintaining the body's acid-base balance means that renal changes can significantly affect respiratory health. This link is evident in CKD patients, who often exhibit reduced respiratory muscle strength and endurance and, in advanced stages, may experience decreased lung function and vocal cord edema due to uremic accumulation, acid-base imbalance, and volume overload [7].

From these respiratory changes in the patient's voice, we propose that the voice can be an auxiliary indicator and demonstrate its potential through some experiments. We collected data from patients with a wide range of kidney health statuses, from stage 1 CKD (CKD 1) to stage 5 (CKD 5), alongside a healthy control group. We crafted data for CKD prediction by extracting extended Geneva Minimalistic Acoustic Parameter Set (eGeMAPS) features from collected voices and combine them with simple demographic data and biometric data from patients, such as age, sex, Body Mass Index (BMI), etc. We train the model with an internal training set and validate it with an external validation set. Consequently, we developed an automated CKD diagnostic system demonstrating an F1 score of 0.86.

This approach could serve as non-invasive and cost-effective markers for the early detection and monitoring of CKD and revolutionize early diagnosis and severity prediction, offering a readily accessible and repeatable tool for tracking disease progression. This could significantly improve patient outcomes by enabling timely interventions while also reducing the reliance on traditional, invasive testing methods.

Our study has opened up new research possibilities by demonstrating the potential of speech-based diagnosis in CKD management, where the interest and importance of remote health monitoring are increasing. The high accuracy of our system in severity detection and simple inspection of CKD through voice analysis can lead to significant improvements in the field. This introduces the feasibility of non-invasive and cost-effective methods for screening severity and presents scenarios for using such techniques in a real-world setting. This development opens new avenues for remote patient monitoring, providing a more accessible and less burdensome alternative to conventional diagnostic methods, and has great potential for improving patient care for CKD and potentially other chronic diseases.

The structure of this paper is organized as follows: Section 2, *Preliminaries* provides a detailed description of the dataset, preprocessing steps, and evaluation metrics used in our study. Section 3, *Methods* elaborates on the methodologies employed for feature extraction and classification, along with the explainable Artificial Intelligence (AI) techniques applied. Section 4, *Results* presents a comprehensive analysis of the experimental results. Finally, Section 5, *Conclusion and Future work* concludes the paper with a summary of findings and discusses potential avenues for future research.

II. Preliminaries

A. Dataset Description

The dataset utilized in this research consists of health information and audio recordings collected under informed consent from patients at Seoul National University Bundang Hospital and Sacred Heart Dongtan Hospital, both highly esteemed institutions in Korea. The data collection targeted individuals presenting with kidney disease, during which each participant recorded six sentences. This set of sentences, including a self-introduction, was carefully developed in collaboration with the Department of Korean Language and Literature at Seoul National University. The design of these sentences aimed to capture the Korean voice's unique characteristics comprehensively.

In this research, 538 individuals aged 20 years and older were included in cohorts. We compiled a comprehensive dataset that included 887 records of hospital visits, including initial consultations and subsequent follow-up appointments. The follow-ups were scheduled quarterly. For analytical purposes, each record of hospital visitation was considered a separate observation.

The demographic breakdown of the cohort is as follows: 52 individuals were identified as not suffering from CKD, while the remaining participants were diagnosed with CKD at various stages of its progression. Specifically, the distribution was 77 individuals at stage 1, 135 at stage 2, 178 at stage 3, 75 at stage 4, and 20 at stage 5 of CKD. The age distribution of the patient cohort predominantly spans from 50 to 80 years, exhibiting a slightly right-skewed normal distribution with a central tendency around the 60s.

The quantitative analysis of hospital visit records yielded the following results: 70 records were categorized as healthy, 104 were identified as Stage 1, 205 as Stage 2, 339 as Stage 3, 141 as Stage 4, and 28 as Stage 5 CKD. Consequently, the analysis showed a ratio of non-critical (comprising normal and Stage 1 and 2 conditions) to critical (encompassing Stages 3 to 5) conditions as 4:6, based on the data derived from hospital visits.

TABLE I
DATASET DESCRIPTION WITH MERGED CELLS.

Stages	# of patients	# of visits	Severity
CKD 0	52	70	Non-critical
CKD 1	77	104	
CKD 2	135	205	
CKD 3	178	339	Critical
CKD 4	75	141	
CKD 5	20	28	

Before its inclusion in our research, the dataset underwent a rigorous de-identification process, ensuring the removal of personal identifiers, such as patients' names and identification numbers. This step was critical in preserving patient confidentiality and adhering to privacy standards. Furthermore, the dataset received approval for international sharing from the Institutional Review Board (IRB), affirming its compliance with ethical standards and regulations for human subjects research. This careful preparation and ethical oversight underscore the dataset's suitability for our study, providing a foundation for reliable and respectful research into kidney disease diagnosis through voice analysis.

B. Preprocessing Details

We separated the record files into sentence units and suppressed noise to minimize external noise intervention. The number of channels and sampling rate were also converted to Mono and 16 kHz, respectively, to exclude intervention due to differences in recording equipment and software. After that, the 88-dimensional features were extracted from each record, and the duration of the utterance was also calculated.

For patient health data, categorical features were encoded. For males, it was encoded as zero and for females as one. For medical history, such as hypertension, diabetes, and so on, it was encoded as one if the patient had the disease and zero if not.

Since age is a critical feature impacting the diagnosis severity, highlighted by its prominent Shapley Additive exPlanations (SHAP) values in our analysis, we categorized the participants into two cohorts based on the median age threshold of 65. We carried out distinct experiments for each group.

Labeling for critical conditions was conducted based on the expertise of a nephrologist to differentiate between insignificant and vital stages of CKD. Patients categorized in stages 0 (healthy), 1, and 2 were deemed non-critical conditions, whereas those in stages 3, 4, and 5 were identified as critical. The classification of each stage was determined by the estimated Glomerular Filtration Rate (eGFR), along with the presence or absence of proteinuria and hematuria. Specifically, stage 0 is characterized by an eGFR greater than 90 $mL/min/1.73m^2$, without proteinuria or hematuria. Stage 1 patients exhibit an eGFR exceeding 90 $mL/min/1.73m^2$, in conjunction with proteinuria or hematuria. Stage 2 encompasses individuals with an eGFR ranging from 60 to 90 $mL/min/1.73m^2$, accompanied by proteinuria or hematuria. Stages 3, 4, and 5 are delineated for eGFR levels less than 60 $mL/min/1.73m^2$, with specific thresholds set at greater than 30 and less than 60, 15 to less than 30, and less than 15 $mL/min/1.73m^2$, respectively.

C. Evaluation Metrics

In this study, we employed precision, recall, F1-score, and Area Under ROC Curve (AUROC) as metrics to evaluate the performance of our model in diagnosing critical stages of disease. Precision is the proportion of patients correctly identified as critical out of all patients the model classified as such. Recall measures the proportion of actual critical patients that the model correctly identifies. Given our goal to facilitate rapid and accurate disease severity identification outside hospital settings—enabling patients with suspected severe conditions to seek hospital care for comprehensive diagnosis and treatment promptly—recall was prioritized as a key metric. To comprehensively assess our model's performance, we also considered the F1 score and AUROC as supplementary metrics, particularly useful in addressing potential data imbalances. The F1 score, representing the harmonic mean of precision and recall, evaluates the model's accuracy in predicting the critical class and its effectiveness in identifying all actual critical cases. Meanwhile, AUROC assesses the model's overall capacity to distinguish between critical and non-critical conditions, offering insights into its performance across varied dataset distributions. The Receiver Operating Characteristics (ROC) curve, in particular, plots the true positive rate against the false positive rate, providing a visual representation of the model's performance. This multifaceted evaluation strategy ensures a balanced understanding of the model's diagnostic capabilities, emphasizing the importance of early and accurate disease detection.

III. Methods

A. eGeMAPS

The extended Geneva Minimalistic Acoustic Parameter Set (eGeMAPS) [8] is a feature set used in various voice research, including speech recognition and disease classification. It contains 88 parameters, including additional descriptors, such as frequencies (bandwidth, Mel-Frequency Cepstral Coefficient (MFCC) 1–4, spectral flux of F1, F2, and F3), energy and amplitude (shimmer, volume of sound, Harmonic to Noise Ratio (HNR)), spectrum (relative energies of alpha ratio, F2, and F3, H1–H2, H1–A3), time (ratio of cloudless peak, average length and standard deviation of meteoric regions, number of continuous meteoric regions per second). In this study, voice features were extracted using these eGeMAPS, and their validity was verified through experiments. This can also be confirmed from the result that the accuracy is improved when eGeMAPS features are used for disease classification using a linear kernel support vector machine and an ensemble classifier [13].

B. Machine Learning Classifier

In this study, we address the challenge of severity classification using high-dimensional, small-scale datasets, including categorical variables. Our approach leverages advanced machine learning classifiers, specifically the Support Vector Classifier (SVC) [2] and eXtreme Gradient Boosting (XGBoost) [9]. The SVC, a specialized form of the Support Vector Machine (SVM), effectively delineates decision boundaries or hyperplanes in a multidimensional space, facilitating accurate data classification. Concurrently, XGBoost employs an ensemble strategy, enhancing predictive accuracy by amalgamating multiple weak predictors, predominantly decision trees, into a cohesive and potent predictive model. These methods were chosen due to their robust capacity to navigate the complexities inherent in limited-sized, high-dimensional datasets [12] [15].

C. Model Explainability

In this study, we utilized Partial Dependence Plots (PDPs) [3] and SHAP [10] values as methodologies to assess the significance and influence of distinct features on prediction outcomes. PDPs were instrumental in illustrating the global effect of selected features on disease classification by delineating how alterations in the values of these features within their observed ranges impact the model's average predictions. This analysis enabled identifying features with substantial predictive power, enhancing our understanding of the model's functionality.

Simultaneously, SHAP values offered a granular, personalized examination by quantifying the contribution of each feature to individual predictions. This nuanced approach was essential for highlighting the specific roles of certain features, such as age, BMI, and F1 frequency, in the classifier's decision-making processes.

The integration of PDP and SHAP analyses provided a comprehensive view of how features affect disease prediction, significantly improving the interpretability of the model. This dual-method analysis confirmed the model's utility in clinical applications and revealed critical insights vital for directing subsequent research initiatives.

IV. Results

A. Correlation between Severity and Voice features

Before initiating the comprehensive experimental phase, we conduct a preliminary analysis to ascertain the correlation between the variables of interest, namely Critical and eGeMAPS features. This involved examining the relationship between critical conceptualization as a binary categorical variable and voice features characterized as numerical variables to assess the significance of their association. To this end, the point-binary correlation coefficient (r_{pb}) [4] and the corresponding p-value were computed.

The formula for the point-biserial correlation coefficient is below where M_1 and M_0 are the mean values on the continuous variable for the two groups defined by the dichotomous variable, s_n is the standard deviation of the continuous variable, n_1 and n_0 are the number of observations in each group of the dichotomous variable, and n is the total number of observations.

$$r_{pb} = \frac{M_1 - M_0}{s_n} \sqrt{\frac{n_1 n_0}{n^2}} \tag{1}$$

The p-value threshold was established at 0.05, facilitating the isolation of instances exhibiting a p-value below this benchmark, thereby affirming the statistical significance of their association with the feature in question. This process identified 72 significant features. Followings are the foremost ten outcomes.

TABLE II
TOP-10 FEATURES CORRELATED WITH TARGET VARIABLE.

Rank	Feature	abs(r_{pb})	p-value
1	**Age**	0.372	0.00
2	**Hypertension**	0.294	0.00
3	**Diabetes**	0.282	0.00
4	**MeanUnvoicedSegmentLength**	0.196	0.00
5	**spectralFlux sma3 stddevNorm**	0.192	0.00
6	**MFCC3 sma3 amean**	0.166	0.00
7	**loudness sma3 stddevNorm**	0.165	0.00
8	**MFCC3V sma3nz amean**	0.162	0.00
9	**loudness sma3 percentile20.0**	0.158	0.00
10	**VoicedSegmentsPerSec**	0.157	0.00

Inspection of the table elucidates that voice-related features predominantly occupy the upper ranks of the absolute correlation metric, each marked by a p-value less than or equal to 0.05, underscoring their statistical significance.

The findings from these analyses substantiate a notable correlation between voice features and Critical, reinforcing our hypothesis's foundational premises.

B. Overall Performance Evaluations

We conducted a series of comparative experiments to investigate the effectiveness of vocal characteristics in the early diagnosis of CKD. Initially, we focused on well-established health indicators such as age, Hypertension (HTN), and Diabetes Mellitus (DM) due to their significant association with CKD, as indicated by high point-binary correlation coefficients. The research has highlighted the independent correlation of factors like older age, increased systolic blood pressure, the prevalence of Type 2 DM, and a longer duration of DM with CKD incidence [11]. Based on this understanding, these indicators could significantly aid in determining CKD severity. To test this, we conducted classification experiments leveraging these health indicators. Following this, we explored the potential of vocal features alone to differentiate CKD severity levels without incorporating additional health information. The results from these experiments suggested that severity classification is feasible using fundamental health indicators, and vocal features alone can achieve comparable results. Our final experiments combined health and vocal information to classify CKD severity. Here, we used two feature sets of health information. The set 'Health I' is the main feature set we used previously: age, HTN, and DM. The other set, 'Health II', includes the previous three, additional information such as sex and BMI that may affect the voice, and additional medical history information collected for the management of patients with CKD, including heart failure, cancer, cardiovascular disease, and cerebrovascular disease.

TABLE III
PERFORMANCE EVALUATION.

Feature Set	Model	Precision	Recall	F1	AUROC
Voice Only	SVC	0.719	0.68	0.698	0.73
Voice Only	XGBoost	0.757	0.731	0.706	0.76
Health I	SVC	0.74	0.686	0.71	0.73
Health I	XGBoost	0.795	0.754	0.743	0.84
Health I & Voice	SVC	0.854	0.777	0.814	0.88
Health I & Voice	XGBoost	**0.876**	0.816	0.826	0.90
Health II & Voice	SVC	0.835	0.83	0.811	0.91
Health II & Voice	XGBoost	**0.876**	**0.882**	**0.857**	**0.92**

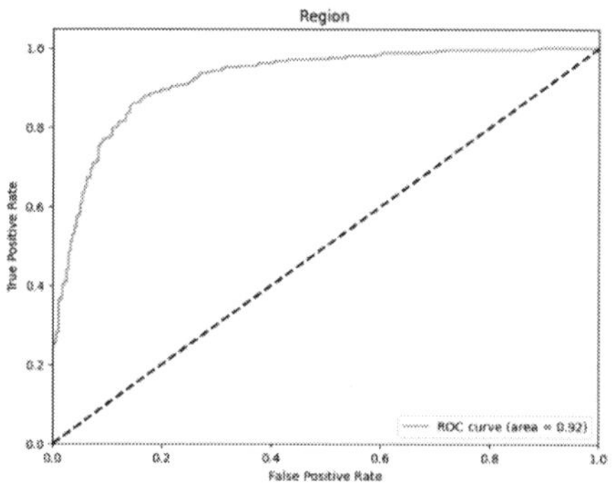

Fig. 1. ROC curve of *Using All*.

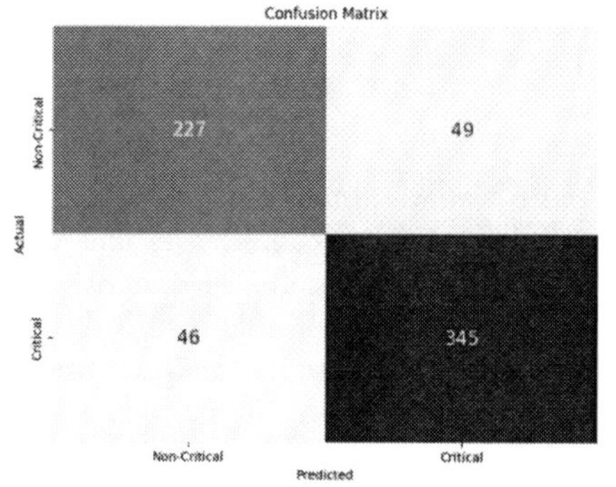

Fig. 2. Confusion Matrix of *Using All*.

The outcomes demonstrated superior classification performance across all metrics when integrating vocal features and all health information, affirming their positive impact on enhancing diagnostic accuracy. The comprehensive results are presented in Table 3, while Figure 1 and Figure 2 depict the ROC curve and the confusion matrix for the 'Using All' scenario, respectively.

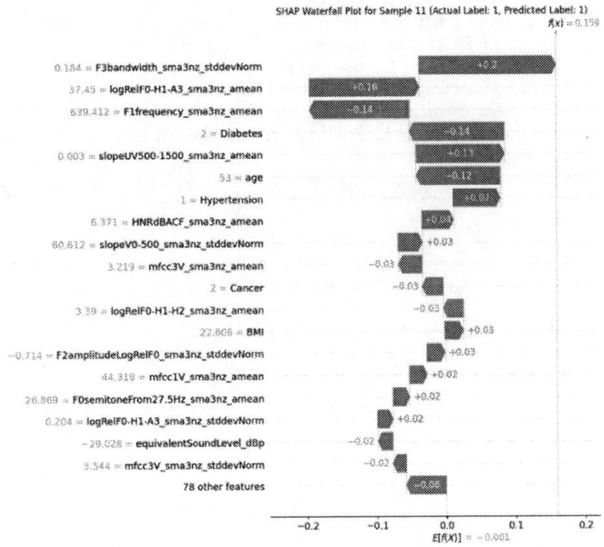

Fig. 3. SHAP Waterfall plot of Critical sample.

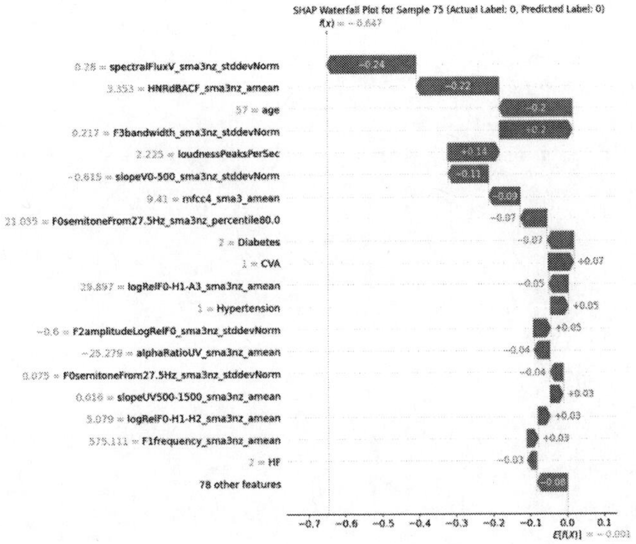

Fig. 4. SHAP Waterfall plot of Non-critical sample.

Figures 3 and 4 illustrate the significance of different features for individual samples, highlighting how certain features contribute to accurately predicting each sample's outcome.

For the first example, where the sample is labeled as 'Critical' (denoted by 1), factors such as age and a history of DM were initially misleading, suggesting a classification as 'Non-critical' (0). However, the presence of voice indicators allowed for the correct classification of this sample. Similarly, the second example, labeled as non-critical, demonstrated that voice features played a crucial role in ensuring the sample was classified accurately despite potential misdirection by some general health indicators. These observations underscore the potential of vocal attributes as reliable supplementary markers for diagnosis, particularly in instances where basic health information alone may lead to ambiguity.

C. Additional Experiments on Age-based separated groups

Age emerged as the most influential factor in our analysis, hinting at a potential overreliance on this variable. To explore this further, we conducted additional tests by dividing the subjects into two groups based on the median age of our patient cohort: those 65 and younger and those older than 65. Each group was then analyzed separately.

TABLE IV
AGE-BASED SEPARATED GROUPS.

Group	CKD 0	CKD 1	CKD 2	CKD 3	CKD 4	CKD 5
age<65	30	66	106	133	59	10
age≥65	21	19	76	197	81	17

The distribution indicates a tendency for the younger group (under 65) to lean towards less severe CKD stages (0, 1, and 2), whereas the older group (over 65) showed a skewed distribution to severe conditions. While the younger group exhibited a relatively balanced distribution across the severity spectrum, the older group displayed a pronounced disparity, with a distribution resembling a 3:7 ratio between non-critical and critical stages, respectively.

To address these disparities and reduce bias, we adjusted our model training approach to utilize Weighted Cross-Entropy (WCE) loss and evaluated the model performance using a Weighted F1-score. This methodology was chosen to lessen the influence of the observed imbalance in disease severity across different age groups. For clarity, we used only XG-Boost, which performed relatively well in previous experiments.

TABLE V
PERFORMANCE EVALUATION ON AGE-BASED SEPARATED GROUPS.

Group	Precision	Recall	F1-Score	AUROC
all ages	0.876	0.882	0.857	0.92
under 65	0.849	0.827	0.852	0.92
over 65	0.909	0.879	0.853	0.92

The results show that the performance did not significantly decrease even when the group was separated based on age. Through this, the model did not depend too much on age and derived classification results by appropriately using the overall information.

V. CONCLUSION AND FUTURE WORK

In this study, we explored the use of patient voice features as biomarkers alongside traditional methods for diagnosing CKD. By employing machine learning techniques, we demonstrated the capability of voice features to classify the severity of CKD accurately. This approach opens new avenues for the continuous and remote monitoring of patients, particularly in severe cases where early detection and swift action are crucial.

Our findings reveal a clear correlation between specific voice features and the severity of CKD, highlighting the potential of vocal analysis in enhancing disease severity classification.

We aim to refine our classification methods by breaking down the severity of CKD into more detailed stages. This endeavor will likely require a more sophisticated experimental setup and the adoption of advanced machine learning technologies, including the potential use of deep learning models known for their robust capabilities. Additionally, we are considering applying voice signals to image-based models, such as multi-channel Convolutional Neural Networks (CNNs), through spectrogram-based voice imaging techniques for more precise voice analysis.

Nevertheless, our study faces the challenge of a limited dataset, a common issue in specialized research areas. To overcome this, we recognize the importance of expanding our dataset comprehensively. Implementing techniques to augment existing data could offer a viable solution to this limitation, enabling more extensive and in-depth research.

REFERENCES

[1] M. Stone, "Cross-validatory choice and assessment of statistical predictions". Journal of the royal statistical society: Series B (Methodological), vol. 36, no. 2, pp. 111-133, 1974.

[2] N. Cristianini and J. Shawe-Taylor, "An introduction to support vector machines and other kernel-based learning methods". Cambridge university press, 2000.

[3] J. H. Friedman, "Greedy function approximation: a gradient boosting machine". Annals of statistics, pp. 1189-1232, 2001.

[4] J. D. Brown, "Point-biserial correlation coefficients". Statistics, vol. 5, no. 3, 12-6, 2001.

[5] A. S. Levey, et al., "National Kidney Foundation practice guidelines for chronic kidney disease: evaluation, classification, and stratification". Annals of internal medicine, vol. 139, no. 2, pp. 137-147, 2003.

[6] V. Pihur, S. Datta, and S. Datta, "Weighted rank aggregation of cluster validation measures: a Monte Carlo cross-entropy approach". Bioinformatics, vol. 23, no. 13, pp. 1607-1615, 2007.

[7] S. Y. Jung, et al., "Voice change in end-stage renal disease patients after hemodialysis: correlation of subjective hoarseness and objective acoustic parameters". Journal of Voice, vol. 28, no. 2, pp. 226-230, 2014.

[8] F. Eyben, et al., "The Geneva minimalistic acoustic parameter set (GeMAPS) for voice research and affective computing". IEEE transactions on affective computing, vol. 7, no. 2, pp. 190-202, 2015.

[9] T. Chen and C. Guestrin, August, "Xgboost: A scalable tree boosting system". In Proceedings of the 22nd acm sigkdd international conference on knowledge discovery and data mining, pp. 785-794, 2016.

[10] S. M. Lundberg and S. I. Lee, "A unified approach to interpreting model predictions". Advances in neural information processing systems, vol. 30, 2017.

[11] S. Damtie, et al., "Chronic kidney disease and associated risk factors assessment among diabetes mellitus patients at a tertiary hospital, Northwest Ethiopia". Ethiopian journal of health sciences, vol. 28, no. 6, 2018.

[12] S. Hegde, S. Shetty, S. Rai, and T. Dodderi, "A survey on machine learning approaches for automatic detection of voice disorders". Journal of Voice, vol. 33, no. 6, 947-e11, 2019.

[13] F. B. Pokorny, et al., "Efficient collection and representation of preverbal data in typical and atypical development". Journal of Nonverbal Behavior, vol. 44, no. 4, pp. 419-436, 2020.

[14] P. Kushner, et al., "Investigating the global prevalence and consequences of undiagnosed stage 3 chronic kidney disease: methods and rationale for the REVEAL-CKD study". Clinical Kidney Journal, vol. 15, no. 4, pp. 738-746, 2022.

[15] M. S. Darouiche, H. El Moubtahij, M. B. Yakhlef, and E. B. Tazi, March, "An automatic voice disorder detection system based on extreme gradient boosting classifier". In 2022 2nd International Conference on Innovative Research in Applied Science, Engineering and Technology (IRASET), pp. 1-5. IEEE, 2022.

[16] S. R. Vaidya and N.R. Aeddula, "Chronic kidney disease". StatPearls. StatPearls Publishing, Treasure Island (FL), 2023.

[17] Centers for Disease Control and Prevention 2022, retrieved April 2024, https://www.cdc.gov/kidneydisease/basics.html

[18] M. Kheirkhahzadeh, "Speech Classification using Acoustic embedding and Large Language Models Applied on Alzheimer's Disease Prediction Task", 2023.

[19] Centers for Disease Control and Prevention 2023, retrieved April 2024, https://www.cdc.gov/nchs/fastats/deaths.html

Assessing Greek National Telemedicine Network

Haralampos Karanikas
Department of Computer Science and Biomedical Informatics
University of Thessaly
Lamia, Greece
karanikas@uth.gr

Vasileios Tsoukas
Department of Computer Science and Biomedical Informatics
University of Thessaly
Dextera Consulting
Lamia, Greece
vtsoukas@uth.gr

Dimitrios Drakopoulos
Dextera Consulting
Athens, Greece
ddrako@dexteraconsulting.com

George Koukoulas
2nd Healthcare Region of Piraeus and Aegean
Piraeus, Greece
koukoulas@2dHR.gov.gr

Angeliki Katsapi
Euro-Mediterranean Institute of Quality and Safety in Healthcare
Athens, Greece
akatsapi@eiqsh.eu

Fotios Rizos
Euro-Mediterranean Institute of Quality and Safety in Healthcare
Athens, Greece
frizos@eiqsh.eu

***Abstract*—This research examines the crucial role of telemedicine in improving healthcare access in Greece, especially on isolated islands and distant mountainous regions. Telemedicine, utilizing Information and Communication Technologies, primarily through interactive videoconferencing, is a significant advancement in the digital health field. It is crucial in guaranteeing fair healthcare access for all individuals, in line with Greece's dedication to protecting the constitutional right to healthcare. This study focuses on the implementation of the Greek National Telemedicine Network (EDIT) and a qualitative assessment of the data recorded by the system. The data analysis of teleconsultation services revealed a clear preference for mental health treatments in both adults and children throughout the study period. Telepsychiatry accounted for the majority of teleconsultations, including over 50% of sessions in the first year and increasing to over 80% from 2017 to 2023. In addition, multiple consultations were conducted to diagnose and treat chronic illnesses, such as diabetes. Over the course of the first five years of operation, the EDIT system had steady annual growth, with an increasing range of examinations being added each year. This signifies the progression of the system and the growing level of approval from users.**

***Keywords*-e-health; telemedicine; Greek national telemedicine network.**

I. INTRODUCTION

In today's digital world, it is crucial to provide technologically advanced solutions that are scalable, cost-effective, and efficient in order to improve the overall health and well-being of individuals [1]. E-health, often known as electronic health, encompasses the utilisation of Information and Communication Technologies (ICT) in the context of healthcare. It includes a wide array of solutions or processes that utilise digital technology to enhance the administration and provision of healthcare. The primary objective of e-health efforts is to enhance the efficiency, cost-effectiveness, and accessibility of healthcare. One of the earliest recorded definitions of eHealth describes it as a developing field that combines medical informatics, public health, and business. It specifically refers to the delivery and improvement of health services and information using the Internet and related technologies [2] [3].

One of the most notable products of e-health could be considered to be Telemedicine. Telemedicine refers to the use of various technologies related to telecommunications and information in the field of healthcare, with interactive video being the most frequently used medium. Various programs were initially developed decades ago. However, the technology has significantly advanced in the previous ten years [4]. Telemedicine offers an appealing substitute for traditional emergency, long-term, and preventive care and can potentially enhance clinical results. Additionally, it is expected to increasingly shift healthcare provision from hospitals or clinics to people's homes in the industrialized world [5].

The technology under investigation has the capacity to enhance traditional healthcare approaches, ultimately ensuring widespread access to high-quality healthcare for individuals worldwide. It may primarily conduct this by enhancing equal access to medical knowledge and facilitating sharing throughout the whole healthcare structure [6]. In light of the statement above, recent research works express that telemedicine has been found to be effective and has beneficial outcomes. These cover the positive impacts on health, improved effectiveness in healthcare delivery, and enhanced technical usability. Other works indicated that it has promise or potential, but further research is necessary to establish definitive findings [7].

Implementing telemedicine in Greece, particularly in isolated islands and rural mountainous regions, is a necessary and practical strategy to fulfill the constitutional mandate of providing equal healthcare access to all citizens, regardless of their location of residence. This work introduces Greece's National Telemedicine Network (EDIT) along with a qualitative research analysis based on various data exported directly from EDIT's central system. The rest of the paper is structured as follows. Section 2 portrays the main concept of the utilization of telemedicine in Greece; Section 3 presents the study's result. Section 4 provides an overview of the current and future directions of the system under consideration. Section 5 offers a discussion, while Section 6 concludes with the findings and future projections.

II. Greece's National Telemedicine Network

The use of telemedicine in Greece especially in islands and remote rural and mountainous inaccessible areas is an ongoing high-importance matter from the Greek Ministry of Health. An endeavor was initiated in 2011 to significantly address the matter, with the following considerations: a) the factors contributing to previous setbacks; b) pertinent studies conducted by the Ministry of Health regarding the advancement of telemedicine in Greece; and c) the implementation of an integrated planning strategy that encompassed not only a technological implementation or telecommunication infrastructure, but also a comprehensive functional framework delineating regulatory framework parameters, institutional coverage, and procedural aspects. In the following paragraphs, a brief description regarding the main system, the subsystems, and the Telemedicine stations is provided.

The National Telemedicine Network - EDIT now comprises the following:

There are 66 Patient Doctor Telemedicine Stations (PDTS) located in Hospitals, Health Centres, and Multipurpose Regional Clinics. These are the actual spaces where the patient is received and where the examination takes place with the attending physician present. They are situated in the country's isolated healthcare facilities, typically Health Centres, Multi-Purpose Regional Medical Centres, and smaller healthcare units.

Twenty-one Consultant Telemedicine Stations (CTS) are located in 12 hospitals of the 2nd HR and the National Emergency Centre (NEC). Additionally, one station has been constructed in Papageorgiou Hospital in Thessaloniki. The CTS functions as the recipient in a telemedicine session. The device is equipped with imaging tools to present the data, vital signs, and images of the patient to the Consultant Physician. It is set up similarly to a PDTS but lacks diagnostic tools and only includes imaging gear. The CTSs are implemented in Regional Hospitals and tertiary hospitals within the 2nd Health Region (HR).

Moreover, 170 Home Care Stations (HCS) are situated in the homes of in-patients or social care facilities inside the 2nd HR international boundaries. Home Health Care Centres are established in the residences of chosen patients to provide direct communication with the Health Unit in their area. The home care and monitoring system is equipped with characteristics that allow it to be used in patients' homes or in collaboration with local social care facilities. Figure 1 illustrates the primary structure of EDIT, while Figure 2 showcases the apparatus found in a telemedicine station.

A. Main Architecture

Telemedicine nodes communicate via broadband networks and specific communication software, exclusively utilizing the Transmission Control Protocol/Internet Protocol (TCP/IP) network protocol and the Multiprotocol Label Switching (MPLS) services of the Public Sector Telecommunications (PST) network. Three logically distinct communication channels are used between the stations for this communication.

- A single channel for transmitting high-definition images and audio for intimate communication between the patient, the patient's doctor, and the consulting doctor.

- A single channel is used to transmit data from diagnostic instruments at the Telemedicine Unit of the PDTS to the Telemedicine Unit of the CTS.

- A secondary channel for transferring additional telemedical data, whether new or old, in digital format that does not come from the PDTS Telemedicine Unit.

The telemedicine network's architecture has specific properties.It enables two forms of telemedicine connections:

(a) Point-to-point (b) Point-to-multipoint (e.g. for tele-education)

Access to the service will be granted exclusively through the telemedicine application and restricted to authorized personnel. Moreover, authorized staff will also conclude the medical appointment via the telemedicine site. The systems will also offer real-time updates on system availability to physicians and authorized personnel.

B. Software

EDIT's software features a versatile and open framework, user-friendly interface, and straightforward functions, providing a distinctive experience for users, including doctors, nurses, other health professionals, and patients. With thorough planning for the operation of the EDIT, it is simple to guarantee access to top-tier health services. A brief overview of some of the software's included features follows:

- Direct medical consultation services are provided by qualified specialist doctors, available to patients regardless of their location or place of hospitalization.
- Reducing needless travel to urban centers to deliver quality health services to patients in rural or island regions.
- Changing the health service delivery model and using innovative methods to get secondary health care to alleviate the strain on outpatient and emergency units in major urban hospitals.
- Implementing telepsychiatry programs.
- Offering ongoing education and job training to medical nurses working in isolated health facilities in collaboration with academic and scientific organizations.
- Ensure equal access to healthcare services for all individuals in the population.

Additionally, it is worth mentioning the key functional characteristics of the software.

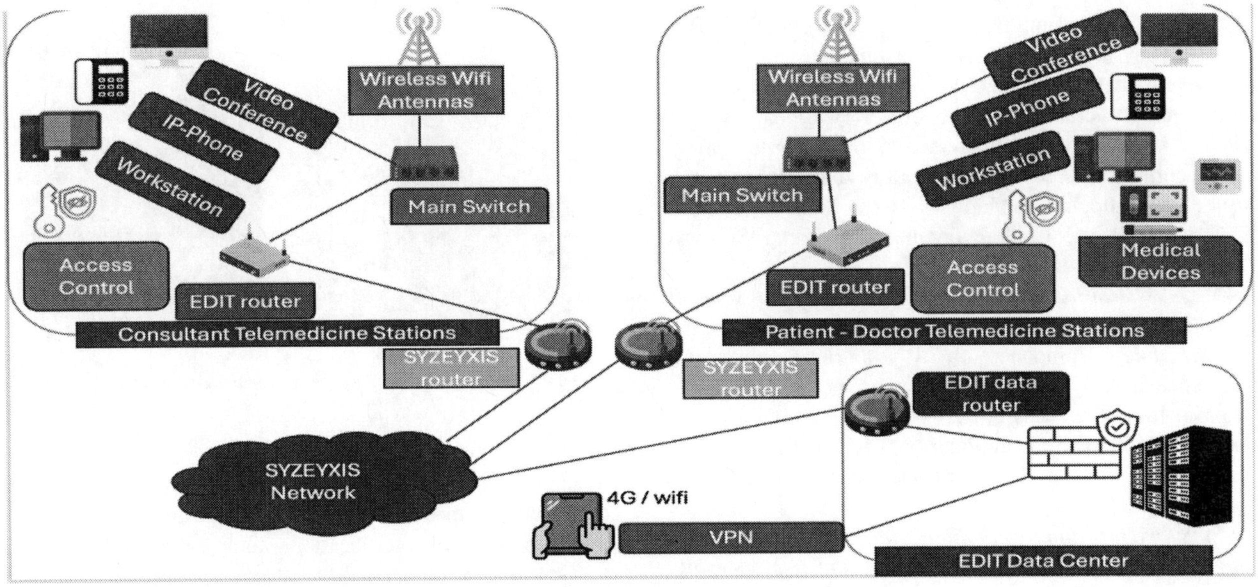

Figure 1. EDIT's Architecture.

Figure 2. A telemedicine station.

1) The application enables remote medical consultations both online and at a later time. Users of the CTS and PDTS platforms can choose to join a live tele-examination or access a patient's uploaded medical data to review later.
2) The application enables communication over the Application Programming Interface (API) and Health Level Seven (HL7). The software provides a well-documented API interface protocol that allows for integration with Electronic medical records (EMRs) and other Hospital Information Systems (HIS) through HL7 and Fast Health Interoperability Resources (FHIR).
3) The software can be configured in High Availability mode to work with an existing Electronic Health Record (EHR) infrastructure. The Carenation application in the current EHR network functions in High Availability mode and will continue to do so with the inclusion of the new EHR health units (PDTS, CTS, HCS).
4) The application combines the functionality of creating a medical tele-appointment, available both online and by scheduling. An option is to seek an immediate tele-appointment for emergencies or scheduled appointments with a certain physician or specialization.

C. Main Components - Sub-systems

1) Subsystem for managing medical devices (Component 1 - Device Gateway): This subsystem facilitates the connecting of medical devices with varying characteristics, measurements, and manufacturers. The medical instruments can be connected through the Device Gateway to transmit the patient's telemetry data, such as video or photos from endoscopes or ultrasound, directly to the medical software on a PC or mobile device and then to a remote site in real-time. A healthcare organization utilizing the EHR system can use current medical instruments for examinations instead of buying new specialized medical equipment due to the interoperability of medical devices.

2) Central Telemedicine Portal Subsystem (Component 2 - Core Care Portal): Appointment management activities are mainly carried out using Component 2 (C2). The system comprises physician and patient management, the patient examination console, auxiliary subsystems for physicians during patient examination, automated examination process flow, and the patient medical record. The system features a central administration and control system that offers enhanced options for configuring the software operation without requiring specialized technical people.

3) Subsystem for Statistics and Analysis (Component 3 - Analytics): Component 3 (C3) was created to provide a detailed analysis of the functionality of telemedicine systems, focusing on both quantitative and qualitative aspects. This subsystem offers pre-designed analytical statistical reports and allows users to create custom reports or export data in formats such as xls, CSV, etc., with precise coding for further analysis. The recording and retrievability of all system data for historical usage are particularly important. C3 can be accessed through C2, which is the Central Telemedicine Portal Subsystem, depending on the user's roles.

4) Subsystem for managing the availability and scheduling of both regular and emergency telemedicine visits (Component 4 - Schedule): Component 4 (C4) is a crucial subsystem for the efficient daily functioning of a telemedicine system. C4 coordinates the process of verifying and certifying the presence of doctors and the availability of telemedicine stations at various appointment sites alongside the central management system.

5) Subsystem interoperability with external systems and applications (Component 5 Integration and Interoperability): Component 5 (C5) was created to enable EDIT's software to work with current or upcoming eHealth subsystems or other systems. It can work with Hospital Information Systems using the HL7 - FHIR protocol, prescription systems, and other systems with specified APIs that support open interfacing. This subsystem is responsible for implementing Single Sign-On (SSO) apps as needed and has an API for connecting with third-party systems.

6) Home care subsystem (Component 6 - Home care environment and functionality).: Component 6 (C6) carries out all functions associated with the Home Care System for Patients. This subsystem allows for the personalized distribution of instructive and educational content. The only subsystem that interacts directly with patients is the one that includes a portal for patients and their aides. An overview of C6's main functionality characteristics is provided below:

- A system designed to provide tailored instructional and advising material for patients.
- Doctors employ both synchronous and asynchronous theta examination in the telemedicine system's second subsystem to offer a uniform interface for all users.
- The system facilitates the transfer of medical examination data from home medical equipment, including oximeters, thermometers, and pressure gauges, through the first subsystem to the Carenation application and subsequently in real-time to telemedicine units as needed.
- Implemented a web-rtc-based video-conferencing system to enable direct connection through the Carenation application with all telemedicine locations.
- The apps of this subsystem have been optimized for optimal productivity and usability on mobile devices such as tablets and mobile phones. - mobile optimization features.

D. Telemedical devices

The Medical Device Management subsystem (Component 1 - Device Gateway) allows for the integration of medical instruments with diverse characteristics, measurements, and manufacturers into the telemedicine program. The examination data from the medical instruments is transmitted through this interface to a remote location, allowing a medical expert to view the examination in real-time and provide advice or opinions. The Device Gateway is a crucial component of the telemedicine program and is responsible for standardizing the telemetry data from connected medical instruments. The data from medical instruments is categorized into channels to provide an interoperability framework based on the data itself rather than on specific manufacturers or products. This approach reduces or eliminates the requirement for parameterization when integrating new medical devices into the system.

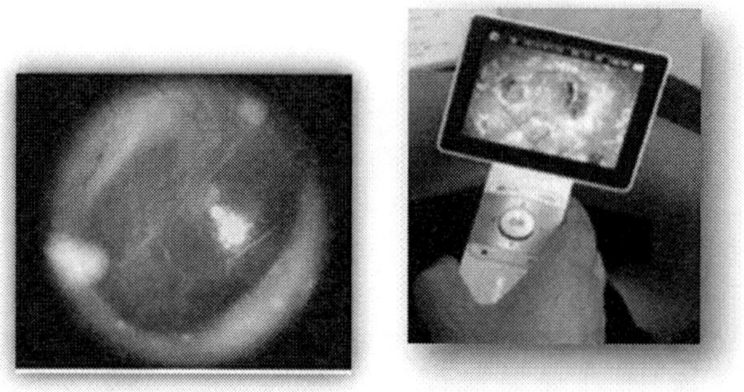

Figure 3. A telemedicine consultation involving an examination camera.

The system supports various devices, such as the following:

- Digital Stethoscope
- Examination Camera
- Dermoscope
- Otoscope
- Ophthalmoscope
- ECG
- Digital Microscope
- Ultrasound
- Medical Cameras for general examination
- Pathology Examination Systems

Figure 3 displays a telemedicine consultation involving an examination camera.

III. Case Study

In order to assess EDIT, a case study was conducted, including the number and the type of teleconsultation for the years 2016-2023. All the data were exported directly from EDIT. The telemedicine services provided by EDIT were primarily implemented in clinical diseases that followed a defined teleconsultation clinical protocol. The implementation of EDIT's telemedicine services did not entail any corresponding organizational adjustments, such as the inclusion of teleconsultation requests in the referral procedures and the introduction of payment. The safety and data privacy concerns were effectively addressed.

A total of 248 teleconsultations were conducted between February 26, 2016 and December 16, 2016. Table 1 categorizes all 248 teleconsultations by speciality.

TABLE I
OCCURRENCES OF DIFFERENT TYPES OF EXAMINATIONS FOR THE YEAR 2016

SPECIALITY	OCCURRENCES
RADIODIAGNOSTICS	1
DERMATOLOGY	2
ENDOCRINOLOGY	1
ODONTIATRICS	109
OPHTHALMOLOGY	2
PEDIATRIC SURGERY	1
TELEPSYCHIATRY WITH CHILDREN	120
SURGERY	1
TELEPSYCHIATRY WITH ADULTS	9
PSYCHOLOGY	2
TOTAL	248

As it could be easily derived from the table shown above, the majority of teleconsultations included telepsychiatry with children at 48.39% and odontiatrics at 43.95%.

The number of teleconsultations in 2017 grew to 352, representing a 41.94% rise. Furthermore, new specialisations such as Hepatology, Diabetology, Pathology, and Social Worker consultation were included for the first time. These additions accounted for 12.5% of all teleconsultations. Telepsychiatry with children accounted for the vast majority of teleconsultations, representing 63.35% of the total. Diabetology followed with a share of 8.8%.

The following year, 2018, once again a new rise in the total number of teleconsultations was observed, with a stunning 239.2%. The examination of teleconsultation services over the aforementioned study period indicates a notable demand for mental health care. Telepsychiatry emerged as the primary area of teleconsultation, with 39.94% of all sessions. Telepsychiatry for children accounted for a significant proportion of teleconsultations for children's mental health, specifically 32.66% of the total consultations. Finally, there was also a substantial portion of diabetology consultations, representing 15.66% of the overall consultations.

In 2019, there were a total of 1638 teleconsultations, representing a 37.19% increase compared to the amount recorded in 2018. 49.81% of the teleconsultations were connected to telepsychiatry, whereas 23.5% were focused on telepsychiatry with children. Moreover, the results indicate a significant 100% rise in psychology consultations and a substantial 317.65% increase in endocrinology consultations.

In 2020, the data on teleconsultations showed a 16.06% increase. Telepsychiatry is the most common form of assessment undertaken with children at 36.29%, followed by telepsychiatry with adults at 35.08%, and psychology at 19.25%. New specialised fields such as Paediatric Developmental Medicine, Urology, Vascular Surgery, Paediatric Allergy, Paediatric Endocrinology, Gynaecology, and General Medicine were introduced.

The data regarding teleconsultations for the year 2021 revealed one more rise in the total of 53.34%. Telepsychiatry with adults is the most prevalent kind of assessment at 27.75%, followed by psychology at 26.58%, and telepsychiatry with children at 26.14%. Furthermore, an analysis of the initial 6 years of operation reveals a steady growth in the number of teleconsultations every year for the years 2017-2021, as shown in Table I.

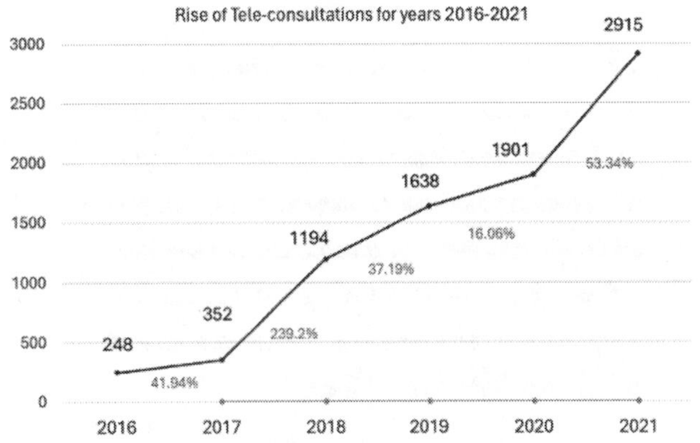

Figure 4. The growth of teleconsultations for years 2016-2021.

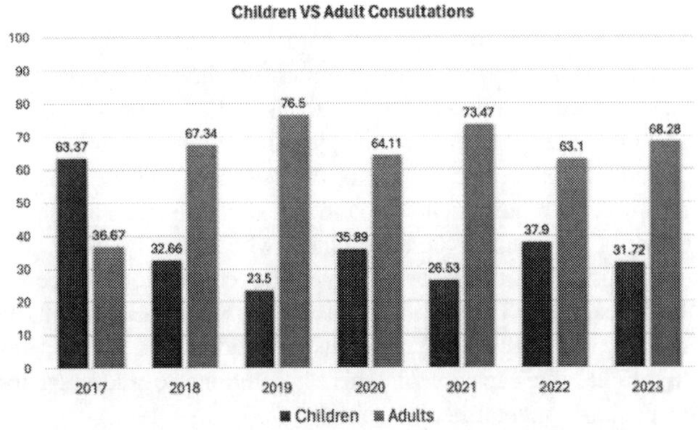

Figure 5. Teleconsultations categorized in regard to whether children or adults were involved.

In 2022, the most common forms of teleconsultations were telepsychiatry with children at 37.58%, telepsychiatry with adults at 28.73%, and psychology at 19.88%.

Finally, for 2023, the most common forms of teleconsultations were telepsychiatry with children at 31.52%, telepsychiatry with adults at 25.55%, and psychology at 22.1%.

In Figure 5, a breakdown of the teleconsultations in regard of whether they were for children or adults is provided, while Figure 6 portrays the types of consultations, whether they were Psychiatry related or not, for the years 2016-2023.

IV. DISCUSSION

The data analysis of teleconsultation services during the study period showed a notable preference for mental health services. Telepsychiatry sessions were the most common type of teleconsultation, representing more than 50% of all sessions in the first year and over 80% from 2017 to 2023. This demonstrates the increasing acceptance and dependence on

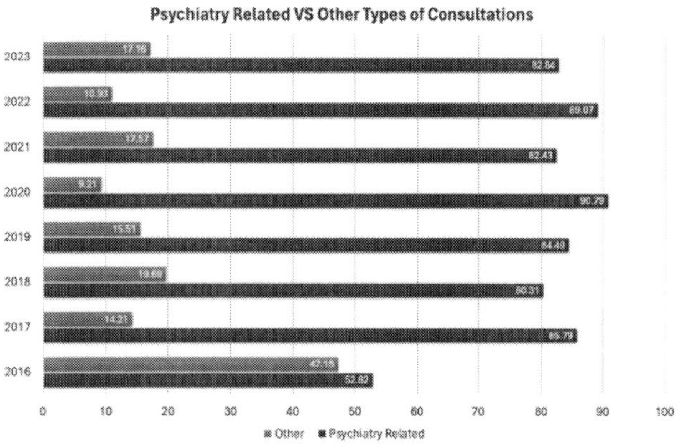

Figure 6. Psychiatry versus other types of consultations for years 2016-2023.

digital platforms for mental health help, indicating a rising awareness and normalization of mental health issues in society.

Following telepsychiatry, teleconsultations for children's mental health, notably in telepsychiatry for children, comprised a considerable portion of the services. This emphasizes the urgent requirement for easily available mental health services for younger demographics, who might be especially susceptible or incapable of accessing conventional face-to-face treatment. The flexibility of teleconsultation platforms in meeting the specific requirements of children and adolescents highlights the capacity of e-health solutions to address gaps in pediatric mental health care.

Diabetology related consultations via telemedicine platforms also represented a substantial portion. This demonstrates a dependence on teleconsultation services to manage chronic illnesses like diabetes, where constant monitoring and regular consultations are crucial. Tele-consultations for chronic disease management show how e-health can improve patient care, treatment adherence, and quality of life for individuals with chronic illnesses.

Furthermore, as derived from Figure 4, Teleconsultations showed steady annual growth from 2017 to 2021. This indicates that the adoption of new and emerging technologies in healthcare is steadily increasing. The technology being studied can improve traditional healthcare methods, ultimately guaranteeing broad access to high-quality healthcare for individuals globally. It accomplishes this by improving fair access to medical knowledge and streamlining its distribution throughout the healthcare system.

Furthermore, new specialties were introduced annually to the current types of consultations. Patients' trust in telemedicine for a variety of examinations is increasing as advancements in software and hardware lead to the development of new medical devices. This progress will make currently unavailable examinations accessible, establishing telemedicine as a primary medium for healthcare.

The distribution of teleconsultations among different specialties demonstrates the adaptability and potential of telemedicine to address various healthcare requirements. It also indicates the changing trends in healthcare delivery towards more convenient and patient-focused solutions. Telemedicine is advancing and provides a significant opportunity to increase healthcare access, particularly in underserved or rural regions, and for people who may encounter obstacles in accessing conventional health services.

V. Current and Future Developments

An ongoing National Programme is currently in place to facilitate the first significant growth of the National Telemedicine Network. The National telemedicine network will be expanded in the 1st, 3rd, 4th, 5th, 6th, and 7th HR. The initiative involves developing a new system that will directly connect with the existing one. Additionally, the existing system will be upgraded in the 2nd HR to include more regional equipment and subscription services. Some of the additions are the following:

1) Three hundred and fifty-five new Patient Doctor Telemedicine Stations - PDTS will be placed in particular Health Facilities nationwide. The PDTS stations are categorized based on space availability data and the operational readiness of each health facility.
2) Thirty-five new Telemedicine Consultant Telemedicine - CTS will be placed in designated Health Facilities. CTS stations are categorized based on space availability data and operational requirements of each Health Facility.
3) Five Telemedicine Training Stations with CTS and PDTS features will serve as training centers for new system users and will be placed in University Hospitals nationwide.
4) Home Monitoring Systems - HCS: 3,000 units with direct communication with the EDIT and related software
5) Medical diagnostic devices for sexually transmitted infections and educational facilities.
6) Three new regional Control Centres and one Command & Control Centre at the Ministry of Health.

In addition, other software changes and subsystems are being prepared for integration into the Central EDIT system. Some of these include but are not limited to, the following.

A new subsystem that focuses on enabling EDIT's healthcare professionals to utilize the globally acknowledged clinical decision support system UpToDate [8].

The nursing staff can utilize the subsystem to access up-to-date medical material for addressing clinical inquiries with dependable, scientifically grounded advice (such as publications, conference papers, best practices, etc.) to enhance patient care and quality of service. This subsystem is crucial for the project as it provides high-quality training content and clinical decision support systems for physicians working in remote NHS units, which is a key measurable goal of the project funded by the Recovery Fund operation.

The subsystem's functioning seeks to enhance the medical services supplied and ensure the availability of high-quality medical clinical information for continued education and training. Improved documentation of expert opinions can help prevent medical errors and reduce complaints about medical practices. Users, including physicians and nursing staff, can access and search through patient records using the EDIT telemedicine program. This activity aims to decrease the time needed to switch environments (e.g., from EMR to web browser) and promote more service usage, leading to increased physician searches.

The quest for sustainable development of the EDIT consists of three main components: economic, social, and environmental issues. Attaining sustainability necessitates a cautious method in developing policies, focusing on certain goals, and rigorously tracking advancements.

To achieve the aforementioned objective, it is essential to create and incorporate a new business intelligence system into the central system. This system will monitor activities and assist in developing new policies based on the continuously observed data.

The EDIT Business Intelligence (BI) system, designed for the National Telemedicine Network, is a pioneering venture focused on utilizing data-driven insights for data processing, performance measurement, and strategy development. Functioning as a central hub, it guarantees that telemedicine operations are both efficient and transparent while also being responsive to stakeholders' needs.

The system will primarily serve the supervision of operational activities, creating usage reports, and developing policies.

The policies will be utilized by the central authority responsible for overseeing the operations and strategic planning located at the Ministry of Health. The Health Regions' administrations and other relevant agencies, such as NEC and Civil Protection will have access to the information.

VI. Conclusion

In the current digital age, it is crucial to have inventive approaches that are expandable, economically viable, and efficient to improve individuals' physical and psychological well-being. E-health refers to the utilization of ICT in the healthcare sector. Telemedicine is a prominent illustration of e-health's accomplishments. Telemedicine uses information and communication technologies to deliver healthcare services, primarily through interactive videoconferencing. Implementing telemedicine in Greece, particularly in isolated islands and remote mountainous regions, is crucial to guaranteeing every citizen's legitimate right to equitable access to medical care regardless of where they reside. The paper introduces the Greek National Telemedicine Network - EDIT and offers a qualitative study using data sourced from EDIT's central database.

Analysis of data from teleconsultation services showed an unambiguous preference for mental health treatments during the time frame under consideration. Telepsychiatry sessions were the most common type of teleconsultation, representing over 50% of all sessions in the first year and rising to over 80% between 2017 and 2023. This trend highlights an increasing trust and dependence on other forms of mental health assistance, indicating a greater acknowledgment and normalization of mental health issues in society. Child telepsychiatry was an important portion of the teleconsultations focused on children's mental health. This highlights the crucial necessity of providing easily available mental health services for young individuals, who can be especially susceptible or incapable of pursuing conventional face-to-face therapy.

A significant portion of consultations focused on diabetology via telemedicine, emphasizing the importance of such treatments for managing chronic illnesses like diabetes. Regular consultations are crucial for ongoing evaluation, and periodic appointments are necessary for optimal illness management. Utilizing teleconsultations for chronic illness management showcases how digital health could enhance medical attention, treatment adherence, and quality of life for individuals with chronic medical problems.

Regarding the EDIT system and its associated data, there was consistent annual growth over the first five years of operation, with an increase in the variety of examinations included each year. This indicates that telemedicine is progressing and offers a valuable chance to enhance healthcare accessibility, especially in underserved or isolated regions, and also for individuals facing barriers to traditional health services.

Acknowledgment

We acknowledge support of this work by the project "National Telemedicine Network (EDIT)" which is co-financed by Greece and the European Union (Recovery and Resilience Fund - Greece 2.0)

References

[1] F. W. Stander and L. E. Van Zyl, "The Talent Development Centre as an Integrated Positive Psychological Leadership Development and Talent Analytics Framework," in Positive Psychological Intervention Design and Protocols for Multi-Cultural Contexts, L. E. Van Zyl and S. Rothmann Sr., Eds., Cham: Springer International Publishing, 2019, pp. 33–56. doi: https://doi.org/10.1007/978-3-030-20020-6_2 10.1007/978 − 3 − 030 − 20020 − 6_2.

[2] M. Stellefson, B. Hanik, B. Chaney, D. Chaney, B. Tennant, and E. A. Chavarria, "eHealth Literacy Among College Students: A Systematic Review With Implications for eHealth Education," *Journal of Medical Internet Research*, vol. 13, no. 4, p. e1703, Dec. 2011, doi: https://doi.org/10.2196/jmir.170310.2196/jmir.1703.

[3] J. Uribe-Toril, J. L. Ruiz-Real, and B. J. Nievas-Soriano, "A Study of eHealth from the Perspective of Social Sciences," *Healthcare*, vol. 9, no. 2, Art. no. 2, Feb. 2021, doi: https://doi.org/10.3390/healthcare902010810.3390/healthcare9020108.

[4] J. Grigsby and J. H. Sanders, "Telemedicine: Where It Is and Where It's Going," *Ann Intern Med*, vol. 129, no. 2, pp. 123–127, Jul. 1998, doi: https://doi.org/10.7326/0003-4819-129-2-199807150-0001210.7326/0003-4819-129-2-199807150-00012.

[5] P. J. Heinzelmann, N. E. Lugn, and J. C. Kvedar, "Telemedicine in the future," *J Telemed Telecare*, vol. 11, no. 8, pp. 384–390, Dec. 2005, doi: https://doi.org/10.1177/1357633X05011008021 0.1177/1357633X0501100802.

[6] N. M. Hjelm, "Benefits and drawbacks of telemedicine," in *Introduction to Telemedicine, second edition*, 2nd ed., CRC Press, 2006.

[7] A. G. Ekeland, A. Bowes, and S. Flottorp, "Effectiveness of telemedicine: A systematic review of reviews," *International Journal of Medical Informatics*, vol. 79, no. 11, pp. 736–771, Nov. 2010, doi: https://doi.org/10.1016/j.ijmedinf.2010.08.00610.1016/j.ijmedinf.2010.08.006.

[8] "UpToDate: Industry-leading clinical decision support." Accessed: Mar. 09, 2024. [Online]. Available: https://www.wolterskluwer.com/en/solutions/uptodate

Work Related Quality of Life and HIS Usability:

An Examination of Human Factors' Impact on Electronic Health Record usability during the Adoption of a New Electronic Health Record System in Norway

[a] Ove Lintvedt[1,2], [b] Espen S. Nordheim[1], [c] Luis Marco-Ruiz[1], [d] Terje Solvoll[1,2], [e] Rune Pedersen[1]

[1]Norwegian Centre for E-health Research,
University Hospital of North Norway,
Tromsø, Norway
[2]Norway Faculty of Nursing and Health Sciences
Nord University
Bodø, Norway

e-mail: [a]ove.lintvedt@ehealthresearch.no, [b]espen.solbakken.nordheim@ehealthresearch.no,
[c]luis.marco.ruiz@ehealthresearch.no, [d]terje.solvoll@ehealthresearch.no, [e]rune.pedersen@ehealthresearch.no

Abstract—**In this paper, we try to determine the influence of human factors, specifically Work-Related Quality of Life (WrQoL), on the usability of a newly implemented Electronic Health Record (EHR) system in Norway. We used the Work-Related Quality of Life questionnaire to measure human factors in clinical staff (physicians, nurses, and others). The National Usability-focused Health Information Systems Scale (NuHISS) questionnaire was used to measure the usability of the new Electronic Health Record. We performed a one-way Analysis of variance (ANOVA) with the NuHISS score as the dependent variable and the WrQoL score as the factor. The results show a significant effect of Work-Related Quality of Life on Electronic Health Record usability ($p < .001$), meaning that work-related quality of life significantly influences the perception of Electronic Health Record usability. The effects vary significantly depending on professional groups, ages, and genders. These findings underscore the importance of considering human factors in the usability and implementation of Electronic Health Record systems. Further research is needed to understand how human factors affect the usability of electronic health record systems.**

Keywords-Electronic Health Record (EHR); usability; Work-Related Quality of Life (WrQoL); National Usability-focused HIS Scale (NuHISS); human factors.

I. INTRODUCTION

In the past two decades, the evolution of Electronic Health Records (EHR) systems has been marked by rapid advancements [1], promising to enhance healthcare delivery efficiency. However, the transformation has shown critical implementation challenges beyond technical innovation. For example, a report from the National Audit Office of Norway on the use of information technology (IT) in hospitals in Norway highlights severe issues [2], including clinician burnout—like in other Western countries [3]. This fatigue is partly attributed to IT solutions that inadequately support clinicians' work processes and time, underscoring the critical need for EHR systems that are both effective and user-friendly [4]. Studies have shown that the lack of usability of EHR systems causes stress in healthcare workers [5]. This stress is not just a byproduct of new systems but is often rooted in the design of the system [6]. In Norway, the government has promoted the development of EHRs that improve information flow [7]. Since 2021, all the Norwegian Health Authority Regions have started implementing a new EHR system. Three health regions have been focusing on transitioning from the Distributed Information and Patient Data System in Hospitals (DIPS) Classic to the DIPS Arena EHR, and one on implementing the Electronically Published Internet Connection (EPIC) system from DocuLive. This study focuses on the transition to DIPS Arena in the Northern-Norway Health Region Authority.

Despite extensive research on the implementation and the effects on user satisfaction of EHR systems [8][9], there still exists a notable gap in our understanding of what affects the EHR usability experience of healthcare professionals. One study found that the usability experience varies by profession and EHR brand [10], and another found that comprehensive adoption and positive work environments enhance nurses' usability and quality of care [11]. Another study aims to develop virtual health records to access EHR data across healthcare levels and improve usability [19].

However, previous studies have often focused on EHR systems' technical and functional aspects and overlooked the sociotechnical interplay between the systems and the individuals using them. Moreover, the transition phase to a new EHR system is a critical period marked by challenges and steep learning curves. More research needs to be conducted in terms of how human factors influence the usability of the new system.

Recognizing such deficiency, our study aims to explore if human factors, such as Work-Related Quality of Life, can contribute to increasing our understanding of how healthcare professionals perceive the usability of newly implemented EHR systems in Northern Norway. Acknowledging the multifaceted nature of usability, which is influenced by a complex interplay of individual factors [20], system design, and organizational structures, it is necessary to explore different human factors that contribute to our perceptions of a new EHR system. The nuanced relationship between technology and user can give us more understanding of the new system beyond the technical and operational needs and allows us to find out if the well-being of the users plays a role in the use of EHR systems.

The two research questions that guide this study are: 1) What is the effect of Work-Related Quality of Life on usability (NUHISS) and 2) Does the impact of these human factors on EHR usability vary according to demographic and professional characteristics?

To achieve these objectives, this study analyzes clinical workers effect of Work-Related Quality of Life (WrQoL) on usability, using the instrument National Usability-focused HIS Scale during the transitioning to new EHR systems. This approach can provide insights into the specific human factors that influence the success of EHR systems' implementation.

The rest of the paper is structured as follows: Section II presents the methods, including setting, data collection, and analysis techniques. Section III presents the results. Section IV discusses the findings. Finally, Section V is the paper's conclusion, with recommendations for future research.

II. METHODS

A. Setting

Norway has organized the governance of the hospital sector under four regional health authorities responsible for South-East, West, Central, and North. In 2021, all these regions were transitioning towards adopting a new Electronic Health Record (EHR) system. Nonetheless, the Northern Norway Regional Health Authority distinguished itself by completing the implementation across its hospitals, shifting from the EHR system DIPS Classic to DIPS Arena. The hospitals included in the study are the University Hospital of North Norway (UNN), Nordland Hospital (NLSH), and Finnmark Hospital (FSH). The selection of the Northern Norway region was based on its position as the only region that had fully implemented the new EHR system, allowing for an examination of the post-implementation of the new system. No other selection guidelines were provided. Subjects of interest had to be selected afterwards. However, the findings need to be contextualized within the characteristics and experiences of Northern Norway's hospitals. The survey was conducted at the end of 2021, following the implementation of the new DIPS Arena EHR system.

B. Data collection

The survey is based on a previously validated questionnaire [14] distributed in 2016 and 2018. In addition to the old survey, this new survey also consists of the survey instruments Work-Related Quality of Life scale (WrQoL) and National Usability-focused HIS Scale (NuHISS). WrQoL [13] has 23 items and a six-factor structure. The instrument is a psychometric scale used to measure human factors that focus on work situations and the quality of life, including factors on General Well-Being, Home-Work Interface, Job and Career Satisfaction, Control at Work, Working Conditions, and Stress at Work. NuHISS [12] consists of 21 items with a six-factor structure. The usability questions include technical quality, Information quality, Ease of use, Benefits, Cross-organizational collaboration, and Internal collaboration. NuHISS was developed for increased knowledge of Health information systems.

Before conducting the new survey, it was piloted through six interviews to get necessary feedback on the quality of the survey. Some changes were made in 2021 to the previous

TABLE I. DATASET, BASELINE AND 2021 DATA

Health Region	Clinical profession			
	Physicians, n	*Nurses, n*	*Other, n*	*Total, n (%)*
FSH	19	18	23	60 (27.1%)
NLSH	20	32	42	94 (42.5%)
UNN	18	32	17	67 (30.3%)
Total	57 (25.8%)	82 (37.1%)	82 (37.1%)	221(100.0%)

questionnaire as it was reported as too time-consuming. We used the survey program LimeSurvey (LimeSurvey GmbH, Hamburg, Germany) to administer the questionnaire. Anonymity was guaranteed to all participants. The survey was dynamically designed to increase relevance for the participants so they would only answer relevant questions. The questionnaire mainly used a 5-point Likert scale ('Completely disagree,' 'Partially disagree,' 'Neutral,' 'Partially agree, 'Completely agree'). Selected items for specific queries were agree/disagree or numeric responses.

The participants were recruited through emails, with each hospital responsible for extending the invitation to all their employees. This method was thought to be the best solution as it used existing administrative structures to facilitate the best possible. To reduce non-responses, a reminder was sent twice between September and December 2021.

The sample consisted of 603 hospital employees, of which 221 participants completed the entire questionnaire, resulting in a 36.5% completion rate. Table I shows the distribution of who answered the survey. These respondents consisted of physicians, nurses, and other professionals who comprise various working groups at the hospital that use the EHR. 25.8% were physicians, 36.2% were nurses, and 38.0% belonged to other professional groups.

The sampling method introduced some limitations. The recruitment method could have been better due to our need for more control over the distribution and limited the answers from the hard-to-contact group. Another possible bias is selection bias, where those most interested in answering such surveys are likely the most enthusiastic about responding. To mitigate these biases, we used strategies such as deploying reminders to improve the response rate.

C. Analysis/statistical methods

Analysis of Variance (ANOVA) was used to determine the dimensions of Work-Related Quality of Life (WrQoL) that significantly influence usability (NuHISS). The significance level was considered p=.05. The statistical software Statistical Package for the Social Sciences (SPSS) 29 (IBM Corp., Armond, NY) was used for the analysis. We proceeded in three steps. Firstly, we analyzed the significance for the overall groups (all age ranges, all professions, genders).

Secondly, we analyzed the significances for each of the groups. Thirdly, we studied the interactions between WrQoL and age, gender, and profession.

A minimal quantity of missing data was observed for the NuHISS variable (n=15, 6.8%), while none were found for WrQoL. Various imputation methodologies have been proposed to address data Missing Completely At Random (MCAR) and in scenarios without systematic patterns of missing data [15]. In this study, we handled missing data by presuming MCAR, as articulated by Little [16]. Our analyses substantiated the MCAR assumption ($\chi2$=.393, df=1, p=.531). Subsequently, we utilized the Expectation Maximization (EM) algorithm within SPSS to impute missing values, which estimates the dataset's means, correlations, and covariances. We adjusted for covariates by including possible confounders and interactions. Age was treated as an ordinal variable, while profession and gender were treated as categorical variables. The scales in WrQoL and NuHISS were treated as ordinal variables.

D. Ethics

The data-protection officer at the University Hospital of North Norway has approved the study. The Regional Committee for Medical and Health Research Ethics Northern-Norway has been consulted, but they concluded that approval was not required.

III. Results

The impact of Work-Related Quality of Life (WrQoL) on the National Usability-focused HIS Scale (NuHISS) scores was examined across various professional groups, age categories, and gender. The ANOVA analysis will show if there are any significant differences in the mean scores of NuHISS across the groups of professions, age, or gender.

A. Baseline data

The number of participants who completed the survey and were EHR users was n=221 (82.5%). Of this group, 70.6% were female. The average years of experience was 17 years (sd=10.9), and the average duration in the current position was 7.6 years (sd=8.1). The mean age of EHR users was 45.7 years (sd=11.6). In terms of professional roles among EHR users, physicians constituted 25.8% (n=57), nurses 37.1% (n=82), and other clinicians also made up 37.1% (n=82). Age distribution within the EHR users was as follows: 10.0% (n=22) were between 18-29 years, 25.3% (n=56) were between 30-39 years, 23.5% (n=52) were between 40-49 years, 28.5% (n=63) were between 50-59 years, and 12.7% (n=28) were 60 years or older.

We checked interaction effects when comparing the groups of WrQoL based on profession, age, and gender. No interactions were found. Lavenes test is the test of homogeneity of variance, checking if the variance in scores is the same for each of the groups. A non-significant result does not violate the homogeneity of variance. Eta squared (Eta2) is used as an effect size measure. It quantifies the strength of the relationship between variables in the ANOVA analysis, quantifying the proportion of variance explained.

TABLE II. ANOVA RESULTS

Factor[a,b]	Subgroup	n	F	df1, df2	p[c]	Eta^2
Profession	All	203	10.441	2, 201	<.001	.093
	Physicians	56	1.313	2, 54	(.277)	.046
	Nurses	79	4.085	2, 77	.021	.096
	Other	68	6.215	2, 66	.003	.158
Gender	All	203	10.636	2, 201	<.001	.063
	Female	143	7.589	2, 140	<.001	.071
	Male	61	3.218	2, 58	.047	.044
	Other	2	-			
Age group	All	204	2.164	14, 189	.006	.074
	-29	21	.186	2, 10	(.833)	.036
	-39	49	.703	2, 15	(.510)	.086
	-49	49	2.172	2, 14	(.151)	.237
	-59	57	5.191	2, 23	.014	.311
	60+	28	2.060	1, 4	(.225)	.340

a. Lavenes test is n.s. for all factors.
b. Interaction effect is n.s. for all factors.
c. Non-significant p in brackets ().

B. Results by professions

By profession, the results show a statistically significant relationship between WrQoL and NuHISS score for all the clinical groups (p < .001); see Table 2. WrQoL for the specific professions, nurses (p = .021) and other professions (p = .003), significantly affect NuHISS. Physicians are the only profession that is not significant in the effect of WrQoL on NuHISS (p = 1.313). We combined physicians and nurses as one clinical group to overcome the small sample size of physicians. For this larger combined group, we have a significant effect for WrQoL on NuHISS (F(2,134) = 4.808, p=.010, Eta^2=.067). All significant results show that low scores on WrQoL relate to lower scores on NuHISS; see Figure 1.

The effect sizes are large (see Eta^2 in Table II) for all professions, as well as for the nurse and the other profession groups. Knowing the level of WrQoL predicts a large amount of total variance in NuHISS. The Eta^2 for the combined group of physicians and nurses has a medium effect.

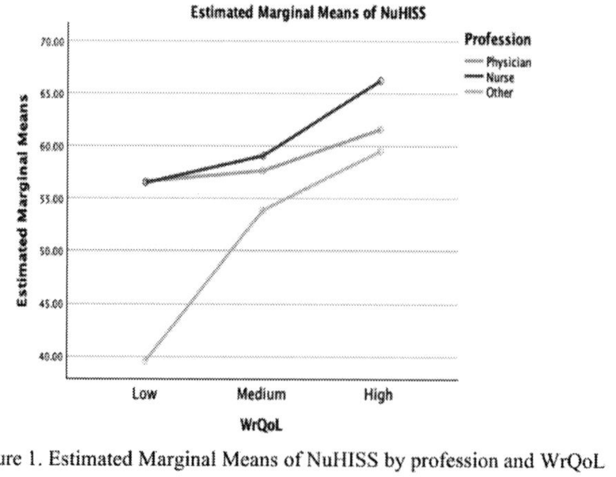

Figure 1. Estimated Marginal Means of NuHISS by profession and WrQoL

C. Results by age

Age shows a statistically significant effect of WrQoL on the NuHISS score for the overall group (all ages combined) ($p = .006$). By age groups, only the groups 50-59 ($p = .014$) were significant. All significant results show that low scores on WrQoL relate to lower scores on NuHISS; see Figure 2. Even if the groups did not reach significance, the trends for each group except the group 18-29 show the tendency that low scores on WrQoL predict low scores on NuHISS.

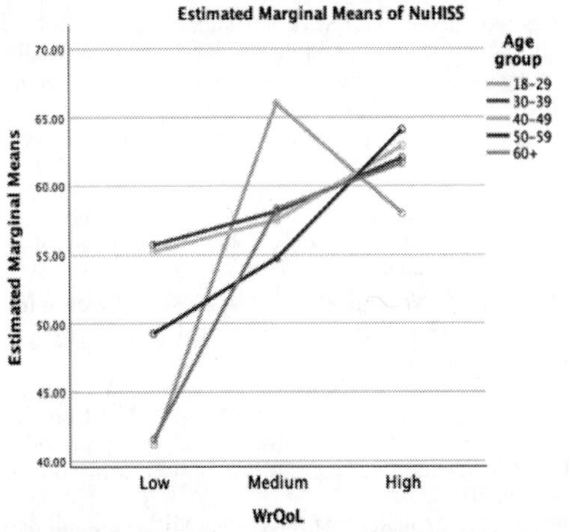

Figure 2. Estimated Marginal Means of NuHISS by age and WrQoL

D. Results by gender

Gender shows a statistically significant effect of WrQoL on NuHISS score for the overall group (all gender) ($p < .001$). When splitting the group by gender, WrQoL for both female and male had a significant effect on NuHISS ($p < .001$ and $p = .047$, respectively). All significant results show that low scores on WrQoL relate to lower scores on NuHISS; see Figure 3.

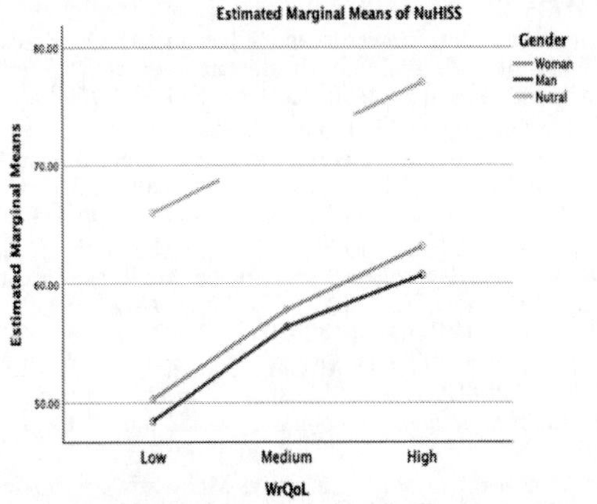

Figure 3. Estimated Marginal Means of NuHISS by Gender and WrQoL

Effect sizes is medium (see Eta^2 in Table 2) for all genders and for female, and low for male. A medium amount of total variance in NuHISS is predictable from knowing the level of WrQoL and gender.

IV. Discussion

Our study sought to explore whether there is a relationship between human factors and how healthcare professionals assess usability during the transition to a new EHR system in Norway. The findings shed some light on this theme and can give insights that can improve our understanding of the implementation process and help adopt new EHR systems.

Our analysis revealed a statistically significant relationship between Work-related Quality of Life (WrQoL) and usability, as measured by National Usability-focused HIS Scale (NuHISS) scores across the combined sample of healthcare professionals. This finding underscores the importance of considering human factors when evaluating EHR system usability. However, a more nuanced exploration could be warranted to understand the specific mechanisms through the subparts of WrQoL and how they influence usability, particularly in sociotechnical aspects and organizational factors.

Furthermore, our study identified notable differences in the impact of WrQoL on NuHISS scores across different professional groups. While nurses and other clinicians showed a significant relationship between WrQoL and NuHISS scores, physicians did not show the same association. This disparity suggests that the influence of human factors on EHR usability may vary depending on specific job roles, task requirements, and organizational contexts. To better understand these differences, future research should explore the underlying factors contributing to variations in the impact of WrQoL on usability among different healthcare professionals. When we combine the clinical groups for physicians and nurses, we have significant results. As the combined sample group is larger than the clinical groups, the non-significant result for the physician group could be due to small samples.

Additionally, our findings highlighted the effects of age and gender on the relationship between WrQoL and NuHISS scores. While the age group 50-59 showed significant results, younger age groups did not show the same patterns. Further investigation into the underlying reasons for these demographic differences is needed to understand why there are differences and ensure a high EHR system usability for all healthcare professionals. The grouping of age groups (5 groups) requires a larger sample to demonstrate if it is also significant for other groups.

By examining the relationship between human factors and usability during the implementation process, we find that human factors play a role in shaping reported EHR system usability. This finding underscores the importance of considering these elements during transitioning to new EHR systems. Our findings need further exploration and should be elucidated in a context within the socio-technical field, looking at how factors, such as context, organizational factors, and human factors, contribute to the socio-technical interplay of implementing a new EHR system.

Despite these insightful findings, our study has limitations. The setting of the study is limited to the hospitals in Northern Norway, and the findings need to be interpreted in that context. Furthermore, breaking down the groups into subgroups makes the sample size small, while necessary for detailed analysis, introduces challenges related to sample size and increases the likelihood of potential biases. The absence of a significant effect in specific subgroups may be attributed to factors such as homogeneity of variances, small sample sizes, minimal differences between group means, considerable within-group variation, non-normality, and outliers. These factors are essential to consider as they can influence the F-value in a one-way ANOVA, affecting the interpretation of the data.

While our study uncovered several significant associations between WrQoL and NuHISS scores across different professional groups, a more in-depth examination is warranted to elucidate these findings' underlying dynamics and implications. Future research could address these limitations by employing follow-up studies or qualitative methodologies [17] to provide a more comprehensive understanding of the complex interplay between human factors and usability with EHR systems. With this regard, implementation research could help us better understand the subjective perspective of EHR users to improve their design [18].

V. Conclusion

The study sought to explore if we could see a relationship between reported Work-Related Quality of Life (WrQoL) and usability, as measured by National Usability-focused HIS Scale (NuHISS) score. Our findings show that there is a significant effect between the variables. Low scores on WrQoL predict low scores on usability. We also show that there are some different effects on different demographic groups and between healthcare professional groups. This should be looked at further to increase our understanding of how human factors influence healthcare professionals' views on Electronic Health Records (EHR) system usability and increase our knowledge of what needs to be considered when implementing a new EHR system. When measuring the usability of a health information system, other factors should be considered, such as organizational factors and access to patient history. In Norway, the patient's journal is not one but separate for each provider, such as hospitals and GPs. A Norwegian project, the Valkyrie project [19], intends to give access to all relevant patient data independent of where the journal is stored, and we want to see how this could impact the evaluation of the same EHR just with access to additional patient history.

References

[1] R. S. Evans, "Electronic Health Records: Then, Now, and in the Future," *Yearb Med Inform*, vol. Suppl 1, no. Suppl 1, pp. S48-61, May 20 2016, doi: 10.15265/IYS-2016-s006.

[2] Riksrevisjonen, "Utilization of IT systems in hospitals". "Utnyttelse av IT-systemer på sykehus" Riksrevisjonen, Document 3:6 (2023–2024), 2023. Retrieved: Feb., 2024. Available from: https://www.riksrevisjonen.no/globalassets/rapporter/no-2023-2024/utnyttelse-av-it-systemer-pa-sykehus.pdf

[3] C. P. West, L. N. Dyrbye, and T. D. Shanafelt, "Physician burnout: contributors, consequences and solutions," *Journal of Internal Medicine*, vol. 283, no. 6, pp. 516-529, 2018.

[4] R. L. Gardner et al., "Physician stress and burnout: the impact of health information technology," *Journal of the American Medical Informatics Association*, vol. 26, no. 2, pp. 106-114, 2019.

[5] T. Vehko et al., "Experienced time pressure and stress: electronic health records usability and information technology competence play a role," *BMC Medical Informatics and Decision Making*, vol. 19, pp. 1-9, 2019.

[6] P. Carayon and P. Hoonakker, "Human factors and usability for health information technology: old and new challenges," *Yearbook of Medical Informatics*, vol. 28, no. 01, pp. 071-077, 2019.

[7] Ministry of Health and Care Services, "One citizen – one Health Record", Whitepaper. no. 9 (2012-2013). "Én innbygger - én journal". *St.Meld. nr. 9* (2012–2013), 2012. Retrieved: Feb., 2024. Available from: https://www.regjeringen.no/no/dokumenter/meld-st-9-20122013/id708609/

[8] O. Lintvedt, E. Nordheim, and R. Pedersen, "Electronic Health Records User Satisfaction," presented at the *eTELEMED 2023*, Venice, Italy, 2023.

[9] R. Pedersen, E. S. Nordheim, O. Lintvedt, A. J. Fagerlund, G.-H. Severinsen, and K. Malm-Nicolaisen, "The Knowledge of Implementation Strategies: Impact of the Installed Base," *Studies in Health Technology and Informatics*, vol. 305, pp. 273-276, 2023.

[10] J. Kaipio, A. Kuusisto, H. Hyppönen, T. Heponiemi, and T. Lääveri, "Physicians' and nurses' experiences on EHR usability: Comparison between the professional groups by employment sector and system brand," *International Journal of Medical Informatics*, vol. 134, p. 104018, 2020.

[11] A. Kutney-Lee, D. M. Sloane, K. H. Bowles, L. R. Burns, and L. H. Aiken, "Electronic health record adoption and nurse reports of usability and quality of care: the role of work environment," *Applied Clinical Informatics*, vol. 10, no. 01, pp. 129-139, 2019.

[12] H. Hyppönen et al., "Developing the national usability-focused health information system scale for physicians: validation study," *Journal of Medical Internet Research*, vol. 21, no. 5, p. e12875, 2019.

[13] D. Van Laar, J. A. Edwards, and S. Easton, "The Work-Related Quality of Life scale for healthcare workers," *Journal of Advanced Nursing*, vol. 60, no. 3, pp. 325-333, 2007.

[14] H. Lærum and A. Faxvaag, "Task-oriented evaluation of electronic medical records systems: development and validation of a questionnaire for physicians," *BMC Medical Informatics and Decision Making*, vol. 4, no. 1, pp. 1-16, 2004.

[15] R. J. Little and D. B. Rubin, "*The analysis of social science data with missing values*," Sociological Methods & Research, vol. 18, no 2-3, pp. 292-326, Nov. 1989, doi: 10.1177/2F0049124189018002004

[16] R. J. Little, "A Test of Missing Completely at Random for Multivariate Data with Missing Values," *Journal of the American Statistical Association*, vol. 83, no. 404, pp. 1198-1202, 1988.

[17] L. Marco-Ruiz et al., "Combining Multivariate Statistics and the Think-Aloud Protocol to Assess Human-Computer Interaction Barriers in Symptom Checkers," *Journal of Biomedical Informatics*, vol. 74, pp. 104-122, 2017.

[18] L. J. Damschroder, D. C. Aron, R. E. Keith, S. R. Kirsh, J. A. Alexander, and J. C. Lowery, "Fostering Implementation of Health Services Research Findings into Practice: A Consolidated Framework for Advancing Implementation Science," *Implementation Science*, vol. 4, pp. 1-15, 2009.

[19] T. G. Solvoll, C. Granja, S. Cassidy, Ø. S. Solvang, and O. Lintvedt, "Valkyrie: A Distributed Service-Oriented Architecture for Coordinated Healthcare Services," presented at the *eTELEMED* 2023, Venice, Italy, 2023.

[20] International Organization for Standardization, "Ergonomics of human-system interaction — Part 11: Usability: Definitions and concepts," ISO 9241-11:2018. [Online]. Retrieved: Feb., 2024. Available from: https://www.iso.org/standard/63500.html.

Evaluation of Hong Kong Medical Students' Knowledge, Attitude, and Intention towards the Use of Telemedicine

Ka Chun Fung
JC School of Public Health and Primary Care
Faculty of Medicine, The Chinese University of Hong Kong
Shatin, Hong Kong
Email: 1155163664@link.cuhk.edu.hk

Abstract— **With the increasing adoption of telemedicine in Hong Kong after the COVID-19 pandemic, increased attention has been given to the future development of telemedicine in both private and public healthcare settings because it plays a role in diversifying the demand in medical support by providing a flexible supply of services. Hence, it is important to understand whether the future doctors are ready for the new trend in the health system. The study examines the knowledge, attitude, belief towards telemedicine, and the intention to use telemedicine of medical students enrolled in their clinical years in Hong Kong. Two research questions have been formulated to understand the mechanism of associating knowledge, attitude, and belief with the intention towards telemedicine. A cross-sectional study was conducted on medical students enrolled in their clinical years in Hong Kong from February to April 2023, in which the subjects were required to complete a questionnaire through Qualtrics after offering informed consent. Convenience sampling was adopted for data collection. 135 invitations were sent online, and 83 valid responses were received. After the data collection, descriptive analysis, bivariate correlation and mediation analyses were conducted. Two main results were identified. First, attitude is a necessary step in developing the connection between knowledge of telemedicine and belief towards telemedicine. Second, belief is not the only factor in developing the connection between attitude and intention towards the use of telemedicine. Based on the results, it was concluded that it is critical to have the medical school curriculum incorporate elements of telemedicine to prepare students to embrace the new practicing mode in Hong Kong.**

Keywords- Telemedicine; Knowledge; Attitude; Intention; Medical students; Hong Kong.

I. INTRODUCTION

The World Medical Association defined telemedicine as the practice of medicine over a distance, including diagnostic judgment and therapeutic treatment [1]. However, it does not fully replace the need for face-to-face medical consultations because remote physical examination and imaging tests are difficult to perform. Therefore, the use of telemedicine has been relatively limited since its debut in 1998 and was mostly employed to manage geriatric outreach patients [2]. Nevertheless, the series of healthcare reforms and technological advancement has encouraged more use of healthcare technology and big data. In 2019, the Medical Council of Hong Kong (MCHK) revised its previous guidelines to introduce recommended ethical requirements for doctors to practice telemedicine [3]. However, the usage of telemedicine did not improve much because doctors tend to be confused about the requirement that teleconsultations must provide a standard of care equivalent to in-person medical practice, which is not an absolute parameter to assess the level of care needed to uphold professionalism.

In the meantime, the Hospital Authority has been keen on designing mobile apps for patient management and health education for the past decade, allowing more people to be aware of telemedicine [4]. Around 13,000 patient consultations and health education initiatives have been completed remotely when specialty and outpatient clinics are not in service since the launch of teleconsultations in Hong Kong public healthcare institutions [5]. In 2022, the Hospital Authority also adopted teleconsultation to relieve the burden of public healthcare institutions while handling the fifth wave of the COVID-19 pandemic. Nevertheless, it does not change the fact that Hong Kong is a conservative adopter and lags significantly behind other regions [6].

In contrast, the use of telemedicine in other parts of the world is very prevalent, and medical students have been exposed to telemedicine in their studies. Studies in Nepal, the United States, and Pakistan showed that most students understand telemedicine well and have telemedicine as part of their clinical learning experience [7][8][9]. In most countries worldwide, medical students have a positive perception of telemedicine and are eager to apply telemedicine in their future careers [7][8][9][10]. Nevertheless, Asian countries tend to have less of a positive attitude towards telemedicine due to the limited development of telemedicine in these regions [6]. In terms of specialties where telemedicine could be applied, over half of the student subjects expect a more extensive use in radiology, teleconsultation, and digital documentation of patient's medical history [11]. Meanwhile, the current training may not be adequate for students to face potential challenges [12].

Telemedicine is a dominant change in the healthcare industry. However, medical education traditionally focuses on theories and face-to-face clinical exposure. Given the global trend for shifting towards the adoption of telemedicine, Hong Kong medical schools have attempted to incorporate telemedicine, particularly during the COVID-19 pandemic. At the University of Hong Kong (HKU), a teaching grant has been applied for developing telemedical applications in clinical teaching and learning in different specialties [13]. Thus, it is important to understand whether future doctors are ready for the new trend in the health system.

As the first study of its kind to date in Hong Kong, this work poses two main research questions:

- How does the knowledge of medical students about telemedicine contribute to the attitude and belief towards telemedicine?
- How does the attitude of medical students about telemedicine contribute to the belief and intention towards adopting telemedicine?

The rest of the paper is structured as follows. In the second section, the hypotheses establishment of the study are illustrated. The hypotheses support the discussion of the methodologies adopted in Section III. In Section IV, the results are illustrated and discussed. The paper concludes with the discussion of implication of the study to various stakeholders in Section VI.

II. HYPOTHESES FORMULATION

Prior to the establishment of the two hypotheses from the two research questions, the definitions and relationships among knowledge, attitude, belief, and intention are discussed.

Knowledge is defined as understanding of an item or individual through experience, communication, or inference based on the organization of meaningful information [14]. In the study, the knowledge would refer to the understanding of the respondents (i.e. the medical students) about the application of telemedicine in Hong Kong.

Attitude refers to the mental disposition individuals have towards certain objects, contexts, or other individuals before attempting to make decisions [15]. There are three components to produce attitude: cognitive content or knowledge received from the experience or external information sources, an affective element from the person creating the attitude, and a tendency for the individual to prepare for action [16]. As telemedicine is a complex concept involving high-order cognitive understanding of the technical competence, the attitude discussed in the study would be about how the subjects feel about the adoption of telemedicine in their foreseeable medical profession development in terms of community-based setting/ clinical setting.

Belief refers to an idea that an individual holds as true [15]. There are various types of beliefs because they depend on how the belief is constructed. For instance, one may develop a belief through personal experience, whereas a belief can also be developed through the socialization with others or based on the norms established in a culture. To be more specific about the belief to be tested, an evaluative belief would be the only type of belief to be tested in the quantitative analysis of the study, which is established from the value and attitude towards the object (i.e. telemedicine). In this context, the belief of medical students towards the effectiveness and efficiency of adopting telemedicine in their foreseeable medical profession development is evaluated.

Intention generally refers to the determination of an individual to act in a certain way [17]. However, intention can be classified from different perspectives. To apply the intention into the study's context, intention refers to the determination of medical students towards the use of telemedicine in the future.

Previous studies concerning telemedicine application around the global community have not explored the mediating effects among the four variables. Yet, health education often makes use of the Knowledge-Attitude-Behavior (KAB) model to explain certain group behavior. With the evaluative belief derived from the attitude towards the telemedicine shaped by the individual, it would be reasonable to develop the following hypothesis (H1).

H1: The relationship between knowledge and belief towards telemedicine is mediated by attitude.

Under rational model, a person's intention of acting in a certain pattern is based on what they believe, so that the intention can be well aligned with belief [18]. With the previous discussion on the formation of belief with the attitude, the second hypothesis (H2) is established.

H2: The relationship between attitude towards telemedicine and intention towards the use of telemedicine is mediated by belief.

By examining the two hypotheses, the project aims to provide insights into how to promote the idea of telemedicine in the Hong Kong medical profession, given that the health system gradually adopts the new mode of practice.

III. METHODOLOGY

A. *Study design and study population*

A cross-sectional study was conducted to collect medical students' opinions concerning telemedicine. In the study, all medical students studying clinical years (years 4-6) in the two medical schools: The Chinese University of Hong Kong (CUHK) and HKU, were the target population. As students mainly get exposed to the clinical setting starting from the fourth year of study in the six-year medical program in both universities and the focus of junior year students is to understand the pre-clinical sciences, medical students in their first, second or third years were excluded from the study. In addition, medical students studying other than the two universities were excluded from the study because the training they receive in a medical school other than CUHK or HKU may not accurately reflect the Hong Kong healthcare setting.

Regarding past studies done in other countries, the non-probability convenience sampling method was adopted. As all medical students studying the clinical years share the common points of being exposed to the clinical setting during their clerkship and specialty training and their sociodemographic background was not a significant factor in the study, it was not necessary to have randomization process to control the factors. Because of the convenience sampling, the response rate varies across different studies. Kong et al. involved soliciting 3500 subjects in the population, and 8.2% of them replied [10]. However, considering the potentially low response rate due to the clash with examination among medical students or other engagements, a ratio of 6% of the population was used. With the use of a 5% margin of error with a 95% level of significance considering 6% of the population of 1500 students, the minimum sample size is 74, which is approximately 5% of the total number of medical students studying years 4-6 in the two medical schools.

B. *Measure*

The study mainly used a questionnaire to collect the students' opinions on telemedicine. In most studies, the questionnaire is divided into three parts: knowledge about telemedicine and exposure to telemedical practice in a clinical setting, the perception of telemedicine, and their preparedness to adopt telemedicine [9]. There is no standardized questionnaire in the studies from other countries. However, the questions mainly involve multiple-choice and Likert Scale ratings on the extent to which the subject agrees with the statements presented. The content of the questionnaires shares high similarities. Hence, the questionnaire comprises five sections with 52 questions in total, in which some of the questions have been adapted from past literature to fit the Hong Kong context, as the healthcare system and medical curriculum are not identical between the areas that the literature studies and Hong Kong. At the same time, it is necessary to identify the various dimensions of the variables so that the results can show a more comprehensive picture. All sections except the section about the participants' demographics have included questions with 7-point Likert scales. The main variables to be measured are knowledge, attitude, belief, and intention.

Six items are used to measure knowledge, nine items are used to evaluate attitude, 15 items are used to measure belief, and five items are used to measure intention.

C. *Data validation*

Prior to the data collection, the questionnaire had been translated to Traditional Chinese and then back to English to confirm the accuracy of the wording adopted in the questions and choices. A small-scale pilot test with 10 participants was conducted to run through the processes and evaluate whether the participants shared the same understanding as the researcher when providing responses to the questionnaire. Based on the feedback obtained, refinement of the questionnaire was carried out, and the revised questionnaire was distributed to the target sample population.

D. *Data Collection*

While some of the studies involved more than one medical school in taking part in the study, the researcher sent out an online questionnaire in the form of Qualtrics and allowed for the receipt of responses in a given period [19][20]. No face-to-face physical data collection session was conducted for the study. After completing the informed consent form, the subjects completed the questionnaire from February to April 2023.

E. *Data Analysis*

Statistical Product and Service Solutions (SPSS) version 26 was used to analyze the data. In case of any missing data on a particular question, the whole set of responses was discarded to maintain data integrity. Descriptive statistics were computed. Bivariate correlations among the variables were examined to see whether any hidden relationships would affect the interpretations. For the variables of "knowledge", "attitude", "belief", and "intention", mediation analyses were conducted to find the associations among them. If the direct effect for the independent variable and dependent variable remained statistically significant with the effect of the mediator, a Sobel test would be conducted to verify whether a partial mediation model exists.

IV. RESULTS AND DISCUSSION

A. *Results*

All the measurements applied are reliable with Cronbach's alpha coefficients beyond 0.7 (see Table I):

TABLE I. CRONBACH'S ALPHA

Measured variables	**Cronbach's alpha**
Knowledge	0.90
Attitude	0.79
Belief	0.81
Intention	0.85

A total of 135 invitations had been delivered, and 83 valid responses were received (61.5% effective response rate). Five incomplete entries were removed to maintain data integrity. The summary of the demographics of the 83 respondents is found in Table II.

TABLE II. DEMOGRAPHIC CHARACTERISTICS OF SAMPLE

Respondents (*N*=83)	
Gender	
Male	79.5%
Female	20.5%
Age (years)	
21-22	16.9%
23-24	53.0%
25-26	26.6%
27-28	3.6%
Education Level	
Secondary school	42.2%
Associate Degree	3.6%
Bachelor's Degree	38.6%
Master's Degree	15.7%
Medical Year	
Year 4	32.5%
Year 5	30.1%
Year 6	37.3%
Medical School	
The Chinese University of Hong Kong	34.9%
The University of Hong Kong	65.1%

The correlation matrix with means, standard deviations, and Pearson's correlation coefficients for the variables measured were presented in Table III.

TABLE III. BIVARATE CORRELATION MATRIX FOR VARIABLES

		Mean	SD	1	2	3	4	5	6	7	8	9
1.	Gender	1.20	.41	--								
2.	Age	23.83	1.46	.20	--							
3.	Educational level	2.28	1.17	.14	.44**	--						
4.	Medical Year	2.05	.84	.078	.36**	-.001	--					
5.	Medical School	1.65	.48	-.129	-.29**	-.54**	-.018	--				
6.	Overall Understanding	4.12	.85	.078	.13	.26*	.24*	-1.74	--			
7.	Attitude towards telemedicine	4.84	1.45	.070	.34**	.47**	.17	-3.23**	.62**	--		
8.	Intention to use telemedicine	5.05	1.57	.061	.28*	.44**	.05	-.46**	.47**	.74**	--	
9.	Belief towards telemedicine	4.70	1.10	.008	.28*	.43**	.20	-.40**	.57**	.76**	.62**	--

N = 83 (listwised), *p < .05, **p < .00.

The association between the independent variable and mediator of the first hypothesis (the positive relationship between attitude towards telemedicine and overall understanding of telemedicine) was supported by the positive correlation (r = .62, p < .01). The association between the mediator and dependent variable in the first hypothesis (the positive relationship of belief towards telemedicine and attitude of medical students towards telemedicine) was also supported by the positive correlation (r = .76, p < .01). For the direct association between the independent variable and dependent variable (the positive relationship of belief towards telemedicine and overall understanding of medical students on telemedicine), it was supported by the positive correlation (r = .57, p < .01).

For the second hypothesis, the association between the mediator and dependent variable (the positive relationship between intention towards the use of telemedicine and belief of medical students towards telemedicine) was also supported by the positive correlation (r = .62, p < .01). The direct association between the independent variable and dependent variable (the positive relationship of intention towards the use of telemedicine and attitude of medical students towards telemedicine) was also supported by the positive correlation (r = .74, p < .01).

The positive correlations provide a basis for further investigation with the mediation regression. Hayes Process Macro Model-4 was used to evaluate the direct effect, indirect effect, and total effect. If the indirect effect is statistically significant while the direct effect becomes non-significant or is significantly reduced, a full mediation is achieved. If the total effect and indirect effect remain statistically significant, a Sobel test would be conducted to determine whether partial mediation is achieved. Otherwise, no mediation is achieved [21].

TABLE IV. REGRESSION ANALYSIS FOR MEDIATION OF MEDICAL STUDENTS' ATTITUDE BETWEEN KNOWLEDGE AND BELIEF TOWARDS TELEMEDICINE

Variables	*B*	CI95%	SEB	β	R²	Δ*R*²
Step 1					.64	**.41****
Constant	1.11	[-2.34, 5.07]	.69			
Knowledge of telemedicine	**1.01****	[.38, .85]	.15	**.59 ** **		
Step 2					.85	**.72****
Constant	**1.57****	[.27, 5.54]	.37			
Knowledge of telemedicine	.11	[-.10, .03]	.10	.09		
Attitude towards telemedicine	**.58****	[.44, .70]	.06	**.76 ** **		

N = 83; *p < .05, ** p< .00

Table IV demonstrates the two-step mediation analyses for the first hypothesis. The second condition of the mediation analysis has been fulfilled, as reflected in the positive bivariate correlation between knowledge and belief (see Table III). Knowledge of telemedicine is positively correlated with the attitude (B = 1.01, p < .001). Then, the mediator (i.e., belief) was introduced along with the independent variable (i.e., knowledge). A statistically significant result (B = .58, p < .001) was achieved. The total effect of the two steps creates a statistically significant result (B = .70, p < .001). Meanwhile, the direct effect between the independent and dependent variables has become non-significant (B = .11, p = .25). Therefore, hypothesis 1 was supported with a full mediation effect.

In addition to the regression analysis, the effects of different types of practical knowledge (exposure to what types of telemedicine) were examined. Among the six sources of telemedicine exposure for participants (social media, peers, family members, mass media, medical curriculum, printed material), the variable "family members" did not have statistical significance. In contrast, the variable "mass media" only led to partial mediation. For the remaining sources, the variable "medical curriculum" had the most substantial effect, leading to full mediation.

TABLE V. REGRESSION ANALYSIS FOR MEDIATION OF MEDICAL STUDENTS' BELIEF BETWEEN ATTITUDE AND INTENTION TOWARDS TELEMEDICINE

Variables	*B*	CI95%	SEB	β	*R*²	Δ*R*²
Step 1					.74	**.55****
Constant	1.13	[-1.52, -.26]	.41			
Attitude towards Telemedicine	**.81****	[.56, .92]	.08	**.74 ** **		
Step 2					.84	**.71****
Constant	**.84****	[.27, 5.54]	.23			
Attitude towards telemedicine	**.39****	[.11, .68]	.14	**.36 ** **		
Belief towards telemedicine	**.65****	[.20, .95]	.18	**.76 ** **		

N = 83; *p < .05, **p< .00

Table V demonstrates the two-step mediation analyses for the second hypothesis. The second condition of the mediation analysis has been fulfilled, as reflected in the positive bivariate correlation between knowledge and belief (see Table III). Attitude towards telemedicine is positively correlated with the belief (B = .81, p < .001). Then, the mediator (i.e., belief) was introduced along with the independent variable (i.e., attitude). There is a statistically significant result ($B = .65$, $p < .001$). The total effect of the two steps creates a statistically significant result ($B = .81$, $p < .001$). However, the direct effect between the independent and dependent variables remains statistically significant ($B = .39$, $p = .00$).

Therefore, a Sobel test was conducted to evaluate whether belief influences the relationship between attitude and intention. The unstandardized beta coefficient (B) and standard error (SEB) were calculated to examine the indirect effect. This satisfied the criteria for a partial mediation effect (Sobel test, $z = 2.21$, $p = 0.03$). Therefore, hypothesis 2 was partially supported.

B. Discussion

Full mediation is achieved in the model representing H1, suggesting attitude is necessary for developing the connection between knowledge (overall understanding) and belief towards telemedicine. Telemedicine is a professional and technical development in the medical field, which involves many cognitive processes from learners to understand the new technology. Therefore, it would be logical to have the flow for the students to be first exposed to the concept of telemedicine before they could develop attitudes and beliefs towards telemedicine, justifying the choice of knowing the independent variable. Apart from the logical deduction, past literature can also prove the association between knowledge and attitude, as the finding from the model is consistent with the results of other studies. Kong et al. identified the relationship between telemedicine exposure, interest, and awareness among medical students in the US, in which students with higher exposure to telemedicine would positively associate with the enhancement of awareness and opinion formation of telemedicine [8]. The awareness induced would provoke a higher interest in the use of telemedicine for clinical management. Moser found that 75% of medical students had a positive attitude towards telemedicine while they had knowledge of telemedicine from media and lectures [22].

Concerning the broad scope of the term "belief", the study would narrow the scope of the variable "belief" to evaluative belief. Bramble defined evaluative belief as "a belief that a particular thing is good (or has value) simpliciter. The good simpliciter makes the world go better rather than worse, impersonally considered" [23]. Evaluation derives from the value and attitude towards the object to be discussed (i.e., telemedicine) because the individuals would rely on their values to develop some standards and parameters to compare their understanding of the object with social norms or their internal mindset. From the positive coefficients of the regression model, it is clear that medical students are more positive in having the belief towards telemedicine when they have a positive attitude and knowledge about the new advancement.

The filter of sources of information medical students receive about telemedicine leads to various degrees of mediation in the first hypothesis. Although students can access a broader range of information in the digital era, information about the advancement of medical technology remains disseminated in credible sources, such as academic studies and clinical cases. Even if students can access the information from other channels (e.g., discussion from family members, mass media), the information received is relatively layman as the general information does not provide much insight on how it would apply to the professional medical practice. Thus, the medical curriculum would be the primary source of information that would influence them to understand the new tools. While some students do not have access to telemedical applications, they desire to incorporate the component into clinical training [12].

On the other hand, partial mediation is achieved in the H2 model, suggesting that belief is not the only factor in developing the connection between attitude and intention towards telemedicine. Even though attitude, belief and intention are interrelated, other variables can contribute to the explanation. Chen et al. studied the attitude, intention and behavior of medical students towards telemedicine, in which they found that other external and objective factors would contribute to the determination of the association [10]. Hsieh et al. examined the attitude and intention of medical students towards the use of telemedicine with the integration of the Theory of Planned Behavior, Self-determination theory and technology acceptance model. Among the complex network of theories, they concluded that attitude and perceived control leads to the intention towards the use of telemedicine, and subjective norm does not have statistical significance on the intention [24]. On some occasions, even though the students are eager to embrace the new practicing mode, they may feel that they are not fully ready for the change. The lack of confidence in adjusting to the new practice may be attributed to inadequate training on the relevant telemedical applications [25]. Meanwhile, the change in healthcare institutions' perspective would be significant in motivating medical students to understand more about telemedicine, even though some students may not have a positive attitude towards telemedicine. Based on the literature findings, there are other external factors that can also contribute to the development of intention, which is not necessarily explained by the individuals' factors such as belief.

In addition to the presence of external influence in complicating the explanatory model, the significant direct effect between attitude and intention is also another key consideration in the partial mediation model. Fishbein and Ajzen introduced the Theory of Reasoned Action (TRA) in 1975. In the theory, the two researchers suggest that an individual's behavior is determined by the behavioral intention, which is directly influenced by their attitude toward the behavior and subjective norm [26]. There is not any role of belief in the association, which can demonstrate the non-necessarily mediating process. Besides, it is not uncommon for human thoughts to be biased with emotions and their preferences. Under the circumstances of emotions and preference being dominant, the individual's intention to behave in a certain manner may not necessarily align with their internally constructed belief [18]. The attitude will override the belief to formulate the intention of action directly, which can explain the statistically significant attitude-intention relationship in the mediation process.

Moreover, telemedicine is a new form of technology to be applied in future medical professions, for which the Technology Acceptance Model (TAM) can be used to explain the behavior. Under TAM, the attitude of individuals directly leads to the intention of accepting and adopting the technology [27]. If the attitude among the population is positive, the individuals have a higher intention of using the technology. This echoes the questionnaire result, where the scores for attitude and intention were 4.84 and 5.05 respectively. Based on TAM, the "belief" component does not have any role in the model, suggesting the intention can be directly explained by attitude without belief serving as a mediator.

Therefore, the higher score in the variable "intention" can be explained by possible external influence and a possible non-rational mindset that skips the belief-matching process, demonstrating a partial rather than full mediation.

C. *Limitations and Future Directions*

While the study provides promising results, there are some limitations throughout the process. First, the study did not include other allied health professions. As clinical management in Hong Kong's healthcare system tries to become more integrative, the opinions of other allied health professionals are equally vital in shaping professional practice. Therefore, further studies can be done to understand the perspectives of other allied health students (e.g., students in pharmacy, nursing, and physiotherapy).

Finally, and importantly, the adoption of telemedicine is currently a dynamic transformation in Hong Kong, and thus the knowledge, attitude, belief, and intention measured may not be identical in the future setting. To better visualize the trend, more studies will need to be done on recently graduated doctors and compared to the findings of the existing study to identify potential trends in the change of attitude, belief, and intention.

V. CONCLUSION

Based on the analyses, it is found that medical teaching and learning are essential for facilitating future doctors to embrace telemedicine in their professional careers. The higher knowledge of telemedicine is highly associated with a positive attitude and belief towards it, suggesting that helping medical students embrace this new form of medical practice would allow for a smoother transition from traditional medical care to integrative care with technology incorporated. At the same time, the positive attitude developed could reduce the tension and negative emotions concerning the change in medical advancement, allowing the city to embrace technological solutions for increasingly complex healthcare problems due to changing sociodemographic factors.

Since the trend of the Hospital Authority and other health institutions in Hong Kong promoting telemedicine is inevitable, medical schools should prepare students at an earlier stage on a gradual basis. For instance, students will be introduced to the concepts and application of telemedicine in pre-clinical years. At the same time, the clinical curriculum should incorporate telemedicine practice in clinical rotations to give students hands-on exposure for future practical application.

VI. IMPLICATIONS

The cooperation from the medical community is equally important. Even though the newer generations of professionals are willing to include telemedicine as part of their practices, the protocols and regulations are not updated well enough to accommodate the trend. Hence, medical lawmakers and managers have to ensure that a timely, complete, and practical protocol, as well as a set of rules, are produced in the foreseeable future to protect the rights of medical professionals in adopting telemedicine. Besides, existing practitioners should also equip themselves with the knowledge and skills of using telemedicine in clinical settings, so that their application experience can serve as practical information for the mentees to pay attention to.

ACKNOWLEDGMENT

I would like to express my deepest gratitude to Dr. Lancelot W. H. Mui, my supervisor during my master's study, for offering guidance and support throughout my write-up.

REFERENCES

[1] The World Medical Association. *WMA - The World Medical Association-WMA statement on Digital Health.* [Online]. 2009 [retrieved Apr. 2024]. Available from: https://www.wma.net/policies-post/wma-statement-on-guiding-principles-for-the-use-of-telehealth-for-the-provision-of-health-care/

[2] Legislative Council of the Hong Kong Special Administrative Region. *Development of telehealth services.* [Online]. 2021 [cited Apr. 2024]. Available from: https://www.legco.gov.hk/research-publications/english/essentials-2021ise14-development-of-telehealth-services.htm

[3] The Medical Society of Hong Kong. *Ethical Guidelines on Practice of Telemedicine.* [Online]. 2019 [cited Apr. 2024]. Available from: https://www.mchk.org.hk/files/PDF_File_Ethical_Guidelines_on_Telemedicine.pdf

[4] The Government of the Hong Kong Special Administrative Region. *LCQ3: Telemedicine services.* [Online]. Feb. 2023 [retrieved May. 2024]. Available from: https://www.info.gov.hk/gia/general/202302/15/P2023021500631.htm

[5] The Government of the Hong Kong Special Administrative Region. *LCQ7: Promoting development of telemedicine.* [Online]. Jul. 2022 [retrieved Apr. 2024]. Available from: https://www.info.gov.hk/gia/general/202207/06/P2022070600446.htm

[6] G. Cheng, F. Wu, and D. Huang. *How COVID-19 is accelerating telemedicine adoption in Asia Pacific.* [Online]. May 2020. [retrieved Apr. 2024]. Available from: https://healthadvancesblog.com/2020/05/08/how-covid-19-is-accelerating-telemedicine-adoption-in-asia-pacific/

[7] S. Kazmi et al., "Nationwide Assessment of Knowledge and Perception in Reinforcing Telemedicine in the Age of COVID-19 Among Medical Students From Pakistan," Frontiers in Public Health, vol. 10, Mar. 2022, doi:10.3389/fpubh.2022.845415

[8] S. S. Kong, A. Azarfar, A. Ashour, C. Atkins, and N. Bhanusali, "Awareness and Attitudes Towards Telemedicine Among Medical Students in the United States," Cureus, vol. 12, pp. e11574, Nov. 2020, doi:10.7759/cureus.11574

[9] B. Kunwar, A. Dhungana, B. Aryal, A. Gaire, A. B. Adhikari, and R. Ojha, "Cross - sectional study on knowledge and attitude of telemedicine in medical students of Nepal," Health Science Reports, vol. 5, pp. e532, Jan. 2022, doi:10.1002/hsr2.532

[10] P. Chen, L. Xiao, Z. Gou, L. Xiang, X. Zhang, and P. Feng, "Telehealth attitudes and use among medical professionals, medical students and patients in China: A cross-sectional survey," International Journal of Medical Informatics, vol. 108, pp. 13-21, Sep. 2017, doi:10.1016/j.ijmedinf.2017.09.009

[11] C. Dockweiler and C. Hornberg, "Knowledge and Attitudes as Influencing Factors For Adopting Health Care Technology Among Medical Students in Germany," Journal of the International Society for Telemedicine and E-health, vol. 2, pp. 64-70, Dec. 2014. [Online]. Available from: https://journals.ukzn.ac.za/index.php/JISfTeH/article/view/78

[12] S. Yaghobian et al. "Knowledge, attitudes and practices of telemedicine education and training of French medical students and residents," Journal of Telemedicine and Telecare, vol 28, pp. 248-257, Jun. 2020, doi:10.1177/1357633x20926829

[13] The University of Hong Kong. *Teaching and Learning, Rising to the Challenge.* [Online]. 2020 [cited May. 2024]. Available from: https://review.hku.hk/teaching-and-learning/

[14] X. Lin, "Review of Knowledge and Knowledge Management Research," American Journal of Industrial and Business Management, vol. 9, pp. 1753-1760, Sep. 2019, doi:10.4236/ajibm.2019.99114

[15] Ministry of Business, Innovation & Employment. *Personal beliefs, values, attitudes and behaviour.* [Online]. [cited Apr. 2024]. Available from: https://www.iaa.govt.nz/for-advisers/adviser-tools/ethics-toolkit/personal-beliefs-values-attitudes-and-behaviour/

[16] K. Park, Textbook of Preventive and Social Medicine. 21st Edition. [Online]. 2011. Available from: http://www.goodreads.com/book/show/16247589-park-s-textbook-of-preventive-and-social-medicine2011

[17] Merriam-Webster. *intention.* [Online]. [cited Apr. 2024]. Available from: https://www.merriam-webster.com/dictionary/intention

[18] C. Nunez, "Requirements of intention in light of belief," Philosphical Studies, vol. 177, pp. 2471-2492, Jun. 2019, doi:10.1007/s11098-019-01321-0

[19] H. N. Abraham et al., "Engaging third-year medical students on their internal medicine clerkship in telehealth during COVID-19," Cureus, vol. 12, pp. e8791, Jun. 2020, doi:10.7759/cureus.8791

[20] P. Malhotra, A. Ramachandran, R. Chauhan, D. Soni, and N. Garg, "Assessment of Knowledge, Perception, and Willingness of using Telemedicine among Medical and Allied Healthcare Students Studying in Private Institutions," Telehealth and Medicine Today, vol. 5, Nov. 2020. [Online]. Available from: https://telehealthandmedicinetoday.com/index.php/journal/article/view/228

[21] *Hayes Process Macro - Model 4 - mediation analysis.* [Online]. 2022 [retrieved Apr. 2024]. Available from: https://researchwithfawad.com/index.php/lp-courses/hayes-process-macro-lecture-series/hayes-process-macro-model-4-mediation-analysis/

[22] P. L. Moser et al., "Acceptance of telemedicine and new media: A survey of austrian medical students," Journal of Telemedicine and Telecare, vol. 9, pp. 273-277, Sep. 2003, doi:10.1258/135763303769211283

[23] B. Bramble, "Evaluative beliefs first," in Oxford Studies in Normative Ethics, vol. 8, M. C. Timmons, Ed. USA: Oxford University Press, pp. 258-273, Nov. 2018, doi:10.1093/oso/9780198828310.003.0013

[24] H. L. Hsieh, J. M. Lai, B. K. Chuang, and C. H. Tsai, "Determinants of Telehealth Continuance Intention: A Multi-Perspective Framework," Healthcare, vol. 10, pp. 2038, Oct. 2022, doi:10.3390/healthcare10102038

[25] G. Franklin et al., "How the COVID - 19 pandemic impacted medical education during the last year of medical school: a class survey," Life, vol. 11, Article 294, Mar. 2021, doi:10.3390/life11040294

[26] M. Fishbein and I. Ajzen, Belief, Attitude, Intention, and Behavior: An Introduction to Theory and Research. Addison-Wesley, Reading, MA, 1975.

[27] R. J. Holden and B. Karsh, "The Technology Acceptance Model: Its past and its future in health care," Journal of Biomedical Informatics, vol. 43, pp. 159-172, Feb. 2010, doi:10.1016/j.jbi.2009.07.002